Cambridge Elements ☰

Elements in the Philosophy of Mathematics
edited by
Penelope Rush
University of Tasmania
Stewart Shapiro
The Ohio State University

LAKATOS AND THE HISTORICAL APPROACH TO PHILOSOPHY OF MATHEMATICS

Donald Gillies
University College London

CAMBRIDGE
UNIVERSITY PRESS

Shaftesbury Road, Cambridge CB2 8EA, United Kingdom

One Liberty Plaza, 20th Floor, New York, NY 10006, USA

477 Williamstown Road, Port Melbourne, VIC 3207, Australia

314–321, 3rd Floor, Plot 3, Splendor Forum, Jasola District Centre,
New Delhi – 110025, India

103 Penang Road, #05–06/07, Visioncrest Commercial, Singapore 238467

Cambridge University Press is part of Cambridge University Press & Assessment,
a department of the University of Cambridge.

We share the University's mission to contribute to society through the pursuit of
education, learning and research at the highest international levels of excellence.

www.cambridge.org
Information on this title: www.cambridge.org/9781009467988

DOI: 10.1017/9781009430609

First published 2023

A catalogue record for this publication is available from the British Library

ISBN 978-1-009-46798-8 Hardback
ISBN 978-1-009-43058-6 Paperback
ISSN 2399-2883 (online)
ISSN 2514-3808 (print)

Lakatos and the Historical Approach to Philosophy of Mathematics

Elements in the Philosophy of Mathematics

DOI: 10.1017/9781009430609
First published online: November 2023

Donald Gillies
University College London
Author for Correspondence: Donald Gillies, donald.gillies@ucl.ac.uk

Abstract: This Element begins by claiming that Imre Lakatos (1922–74) in his famous paper 'Proofs and Refutations' (1963–4) was the first to introduce the historical approach to philosophy of mathematics. Section 2 gives a detailed analysis of Lakatos' ideas on the philosophy of mathematics. Lakatos died at the age of only fifty-one, and at the time of his death had plans to continue his work on philosophy of mathematics, which was never carried out. However, Lakatos' historical approach to philosophy of mathematics was taken up by other researchers in the field, and Sections 3 and 4 of the Element give an account of how they developed this approach. Then, Section 5 gives an overview of what has been achieved so far by the historical approach to philosophy of mathematics and considers what its prospects for the future might be.

Keywords: Lakatos, proofs and refutations, revolutions in mathematics, heuristics, discovery, axiomatic method

ISBNs: 9781009467988 (HB), 9781009430586 (PB), 9781009430609 (OC)
ISSNs: 2399-2883 (online), 2514-3808 (print)

Contents

1 Introduction

Imre Lakatos was born in Debrecen in Hungary on 9 November 1922. He graduated from the University of Debrecen in 1944, having studied Mathematics, Physics, and Philosophy. He went on to take a PhD also from the University of Debrecen in 1948. The title of his thesis was 'On the Sociology of Concept Formation in the Natural Sciences'. At the time, the most famous philosopher in Hungary was György Lukács, and Lakatos attended his weekly Wednesday afternoon private seminars. From Lukács, he would have become familiar with a Hegelian version of Marxism. Lukács himself applied this approach to literature, and Lakatos did in fact write some early papers on literature. However, as we see from the title of his Debrecen thesis, he had already by then decided to focus on the natural sciences rather than literature. Later of course he extended the material he considered to include mathematics. This shift away from literature is quite understandable given his undergraduate studies at the University of Debrecen. Moreover, by applying the approach of Lukács to science and mathematics, he was exploring new ground.

In 1956, the Hungarian Uprising occurred, and Lakatos fled to the West with his second wife (Eva Pap) and her father. He obtained a fellowship from the Rockefeller foundation to work for a second PhD at King's College Cambridge with Professor Richard Braithwaite as supervisor. He was awarded a Cambridge PhD in 1961. While working on his PhD he got to know and became very friendly with Popper. In 1960, he joined Popper's department at the London School of Economics as a lecturer. In the 1960s, Lakatos' work became internationally famous, and he was promoted to full Professor in 1969. Sadly, he did not live long to enjoy his success. From 1970, he became increasingly ill and died of a heart attack on 2 February 1974 at the age of only 51.

Lakatos carried out research in the philosophy of mathematics and the philosophy of science. In this Element, I will concentrate on his work in philosophy of mathematics, but I will say just a little about what he did in philosophy of science, since there were important connections between his research in the two areas. Section 2 of the Element deals with Lakatos' own contribution to the philosophy of mathematics. At the time of his death, Lakatos was planning to carry out further researches in the philosophy of mathematics, and perhaps would have modified some of his earlier views, but this was not to be. However, his approach to philosophy of mathematics was taken up and developed by quite a number of researchers in the field in the period from 1975, the year after Lakatos' death, to the present (2023). Sections 3 and 4 of the Element give a sketch of this legacy of Lakatos in the philosophy of mathematics. What is striking is the variety of different ways in which Lakatos' ideas have been developed, not all of which would have been approved by Lakatos himself. In Section 5, I make a few concluding remarks about the character of these developments and whether further progress might be possible in this general approach.

2 Lakatos' Contribution to the Philosophy of Mathematics

My main thesis is that Lakatos' very important contribution consisted in his introduction in his 1963–4 paper 'Proofs and Refutations' of the *historical approach* to the philosophy of mathematics. The striking nature of this paper and its interesting results led other researchers in philosophy of mathematics to take up the historical approach and it was in subsequent years strongly developed, although it had never been used by philosophers of mathematics before Lakatos. In some ways, it is strange and surprising that the historical approach to philosophy of mathematics was introduced as late as 1963–4, because the historical approach to philosophy of science had been introduced in 1840, 123 years earlier, and by 1963–4 had become a very well-established approach to philosophy of science. I will next give a brief sketch of the development of the historical approach to the philosophy of science since this will illustrate the nature of the historical approach and how it differs from other approaches. It will also be helpful because, when the historical approach to philosophy of mathematics was introduced, the way in which it was developed was strongly influenced by the earlier results of the historical approach as applied to philosophy of science.

2.1 The Historical Approach to the Philosophy of Science

The historical approach to philosophy of science was introduced by William Whewell in his 1840 book: *The Philosophy of the Inductive Sciences Founded upon Their History*. The title here gives one of the fundamental ideas of the historical approach, namely that philosophy of science should be based on a study of the history of science. Whewell was well-placed to adopt this approach, since in 1837, he had published a book on the history of science with the title: *History of the Inductive Sciences from the Earliest to the Present Times*. The historical approach can be contrasted to the logical approach adopted by the Vienna Circle. The logical approach to philosophy of science consists in the logical analysis of contemporary scientific theories with little or no consideration of the history of science.

Following its introduction by Whewell, the historical approach was taken up by several philosophers of science and a number of distinguished books using this approach were produced in the next hundred years or so. I will here mention just three notable works. The first is Ernst Mach's 1883 *The Science of Mechanics: A Critical and Historical Account of Its Development*. As well as tracing the history of mechanics, Mach develops some philosophical ideas. Of particular importance are his operationalist account of the concept of mass, and his critique of Newton's idea of absolute space. Instead of introducing absolute space, Mach thought that we should use the frame of reference provided by the fixed stars. These philosophical ideas of Mach's were carefully studied by Einstein, who used them in his development of relativity theory.

The second book I will mention is Pierre Duhem's 1904–5 *The Aim and Structure of Physical Theory*. This introduces a number of philosophical views on science which have remained of great importance ever since. Perhaps the most famous is the *Duhem Thesis*, later extended to become the *Duhem–Quine Thesis*. Duhem employs to perfection the typical technique of the historical approach which consists in using historical case-studies to develop philosophical ideas. The case-study both illustrates and supports a particular view, as well as undermining alternatives. An example is Duhem's critique of inductivism, which is based on a study of Newton's attempt to derive his theory of gravity by induction from Kepler's laws. Duhem argues that this attempt was a failure since, strictly speaking, Kepler's laws contradict Newton's theory of gravity. It is not therefore possible to derive Newton's theory from Kepler's laws by any kind of logical process. Duhem also shows the importance of metaphysical ideas in the development of science and carries out an analysis of scientific discovery based on a case study of the discovery of universal gravitation.

The third book is Karl Popper's 1963 *Conjectures and Refutations. The Growth of Scientific Knowledge*. This was published in the same year as Lakatos's 'Proofs and Refutations', but the articles of which Popper's book was composed had mainly appeared earlier and would have been known to Lakatos while he was developing his new approach to the philosophy of mathematics. The phrase 'The Growth of Scientific Knowledge' in the title of Popper's book illustrates another feature of the historical approach to philosophy of science. This concentrates on the growth of science and tries to discover patterns in this growth. In the case of Popper, this pattern was 'Conjectures and Refutations'.

We see from these examples that, between 1840 and 1963, the historical approach to philosophy of science had been a vigorous tradition producing many notable works. Why then did no one adopt the historical approach to philosophy of mathematics during this period? This is a tricky question, but I will now try to suggest some tentative answers.

One factor was perhaps the tendency to regard mathematics as consisting of eternal, timeless truths, such as $5 + 7 = 12$. Now who discovered this particular truth? It must have been one or more individuals living at the tribal stage before the rise of urban civilisation and writing. But does it really matter, when, where, or by whom, the truth that $5 + 7 = 12$ was discovered? Once it is formulated, we can recognize it as correct and hence its historical origins do not seem to be of great relevance.

A second factor is really a special case of the first. Anyone like Whewell in 1840 trying to construct a philosophy of science had to take account of the changes brought about the Copernican revolution. The ancient Greeks accepted the Earth-centred Ptolemaic astronomy and Aristotle's mechanics. Yet both Ptolemaic astronomy and Aristotle's mechanics came to be rejected in the sixteenth and seventeenth centuries in favour of a Sun-centred astronomy and Newtonian mechanics. This at once gives a picture of science in which older erroneous theories are rejected in favour of newer and better theories, and such a picture of science involves history. By contrast, ancient Greek mathematics was still accepted by nearly all mathematicians as completely correct right down to 1840. This again suggested that there was no need to take an historical approach to mathematics.

One might have thought that the general acceptance of non-Euclidean geometry in the 1860s would have stimulated the development of an historical approach to mathematics. But this did not occur. Many continued to think that non-Euclidean geometry was just a logical possibility and that the true geometry of space was and would always remain Euclidean geometry. An example of this is Frege, who wrote in 1918–19, p. 363:

> Thus for example the thought we have expressed in the Pythagorean theorem
> is timelessly true, true independently of whether anyone takes it to be true. . . .
> It is not true only from the time when it is discovered.

It is worth noting that in 1918–19, Riemannian geometry was well known to mathematicians and had been applied to physics by Einstein in his General Relativity.

Not everyone was so cavalier about non-Euclidean geometry. Some, perhaps the majority, took it more seriously, and argued that mathematics should be founded not on geometry but on arithmetic, through the arithmetization of analysis. Arithmetic provided the requisite eternal and timeless truths. This arithmetical approach led in turn to the three main schools of philosophy of mathematics: logicism, formalism, and intuitionism. These schools are entirely non-historical. Their philosophical views of mathematics are developed without taking the history of mathematics into consideration or even mentioning it at all.

What is particularly remarkable is that Pierre Duhem, one of the leading figures of the historical approach to philosophy of science, actually says explicitly that this historical approach should not be applied to mathematics. He writes in the section of his book devoted to discovery in physics (1904–5, p. 269):

> This importance which the history of the methods by which discoveries are made acquires in the study of physics is an additional mark of the difference between physics and geometry.
>
> In geometry, where the clarity of deductive method is fused directly with the self-evidence of common sense, instruction can be offered in a completely logical manner. It is enough for a postulate to be stated for a student to grasp immediately the data of common-sense knowledge that such a judgment condenses; he does not need to know the road by which this postulate has penetrated into science. The history of mathematics is, of course, a legitimate object of curiosity, but it is not essential to the understanding of mathematics.
>
> It is not the same with physics.

The second two paragraphs of this passage from Duhem are quoted by Lakatos in his 1978b, p. 43, Footnote 3, as an illustration of his claim that:

> The history of mathematics has been distorted by false philosophies even more than has the history of science. It is still regarded by many as an accumulation of eternal truths; . . .

Given then that there was so much opposition to the historical approach to philosophy of mathematics, how was Lakatos able successfully to introduce such an approach in 1963–4? The answer is, I think, quite simple. By the 1950s, the three main, non-historical, schools of philosophy of mathematics were all facing very great difficulties. Logicism and formalism had suffered a damaging blow, some would say an outright refutation, from Gödel's discovery of his incompleteness theorems which were published in 1931. Intuitionism differed in an important respect from logicism and formalism. Logicism and formalism were both attempts to justify the standard mathematics of the time. Intuitionism criticized standard mathematics and sought to replace it by a new constructive mathematics. However, by the 1950s, it had become clear that very few mathematicians were prepared to change standard mathematics for this new constructive version. Thus, intuitionism was implicitly rejected by practising mathematicians.

Naturally, these problems did not lead to the immediate demise of the three schools of philosophy of mathematics. On the contrary, there were in the 1950s, and still are, neo-logicist and neo-formalist approaches to the philosophy of mathematics, while many different versions of constructive mathematics have been developed. Still by the 1950s, it began to look as if the three standard approaches were unlikely to yield any very interesting developments in the philosophy of mathematics. A new approach seemed to be needed, and this was what Lakatos supplied with his 1963–4 paper 'Proofs and Refutations'. I now turn to an analysis of this paper.

2.2 Lakatos' 1963–4 Paper 'Proofs and Refutations'

The paper is stated to be: '*For George Pólya's 75th and Karl Popper's 60th Birthday*' and Lakatos writes that (1963–4, I, p. 1): 'The paper should be seen against the background of Pólya's revival of mathematical heuristic, and of Popper's critical philosophy'. This is certainly true. 'Proofs and Refutations' is modelled on Popper's 'Conjectures and Refutations'. The first version of the paper was read at Popper's seminar in March 1959, and the paper is extracted from Lakatos's PhD thesis of 1961, which had the title: *Essays in the Logic of Mathematical Discovery*, a title which is clearly based on Popper's *The Logic of Scientific Discovery*. In effect Lakatos's strategy was to apply ideas which Popper had developed for science to mathematics. Following Pólya, Lakatos often considers heuristics. Lakatos was also influenced by his Hungarian friend Árpad Szabó, a historian of ancient Greek mathematics, who had been one of Lakatos' teachers in the University of Debrecen, and to whom Lakatos refers on several occasions.

Lakatos begins his paper with a polemic against what he calls 'formalism'. Usually 'formalism' is used to denote the views of Hilbert and his followers, but Lakatos uses it in a wider sense to include also the logicism of Frege, Russell, and the Vienna Circle. Lakatos writes (1963–4, I, p. 3, 1976, pp. 1–2)[1]:

> Formalism disconnects the history of mathematics from the philosophy of mathematics, since, according to the formalist concept of mathematics, there is no history of mathematics proper.

The first part of this passage (before the word 'since') is undoubtedly correct, but the second part contains a rather characteristic Lakatosian exaggeration. Though Lakatos does not mention this, the first part applies also to the third major contemporary school in the philosophy of mathematics, Brouwer's intuitionism, even though Brouwer was not a formalist.

Lakatos is strongly opposed to the disconnection of the history of mathematics from the philosophy of mathematics and repudiates it in a paraphrase of Kant. He writes (1963–4, I, p. 3, 1976, p. 2):

---

[1] In quoting from 'Proofs and Refutations', I will give the page numbers of both the original 1963–4 version published in *The British Journal for the Philosophy of Science* and the book version published in 1976. The book version contains some additional material from Lakatos's PhD thesis, and I will reference these additions just by their page number in the 1976 book, though it should be remembered that they were written in 1961.

> The philosophy of mathematics, turning its back on the most intriguing phenomena in the history of mathematics, has become *empty*.

Here again there is some Lakatosian exaggeration. Logicism, formalism, and intuitionism were all three in great difficulties, but to describe them as empty is hardly fair. They had all three contributed quite a number of interesting ideas and techniques which had proved useful in mathematics.[2] Still the problems in all three approaches certainly justified trying a new approach – the historical approach.

Lakatos goes on to employ a typical method of the historical approach by describing a case history from the development of mathematics and drawing philosophical conclusions from it. However, his use of this standard method has an unusual feature, namely that his paper takes the form of a lively dialogue. He explains the reason for this as follows (1963–4, I, p. 7, 1976, p. 5):

> The dialogue form should reflect the dialectic of the story; it is meant to contain a sort of *rationally reconstructed or 'distilled'* history. *The real history will chime in in the footnotes, most of which are to be taken, therefore, as an organic part of the essay.*

In my account of 'Proof and Refutations' I will rely mainly on the real history in the footnotes; but I will say something about Lakatos' concept of 'rationally reconstructed history' later in the Element (in Section 3.4).

Lakatos' choice of case history was suggested to him by Pólya, and it should be stressed that it concerns a development which is quite central to the history of Western mathematics. It deals with polyhedra, and these are the subject of the final three books of Euclid's *Elements* (Books XI, XII, and XIII). As Lakatos points out (1963–4, I, p. 16, Footnote 1, 1976, p. 14, Footnote 1), Euclid begins by defining specific polyhedra, namely (Elements, XI, pp. 301–2): pyramid, prism, octahedron, icosahedron, and dodecahedron. However, he does occasionally use the general term polyhedron, or rather in Heath's translation, the term polyhedral solid. This occurs for example in Elements, XII, Proposition 17, p. 363.

The final book of Euclid's Elements (XIII) deals with the regular solids. Euclid shows how to construct the tetrahedron (Proposition 13, p. 381), the octahedron (Proposition 14, p. 383), the cube (Proposition 15, p. 385), the icosahedron (Proposition 16, p. 385), and the dodecahedron (Proposition 17, p. 389). Euclid concludes the Elements by showing that these five are the only possible regular solids.

These five regular solids are often known as the Platonic solids, because Plato in the *Timaeus* associates each of the four elements of ancient Greek science

---

[2] I will say a bit more about this in Section 4.1.

with one of the regular solids. Earth is associated with the cube, Air with the octahedron, Water with the icosahedron, and Fire with the tetrahedron. Plato also hints that the heavens might be constructed from the fifth regular solid – the dodecahedron. This leads Popper to claim that (1963, p. 88):

> Euclid's geometry was not intended as an exercise in pure geometry ... but as an *organon* of a *theory of the world*. ... the 'Elements' is not a 'textbook of geometry' but an attempt to solve systematically the main problems of Plato's cosmology.[3]

Kepler at the end of the sixteenth century revived the attempt to use the five regular solids to give an explanation of the structure of the cosmos, though his approach was different from Plato's. His speculations may, however, have stimulated an interest in polyhedra among mathematicians in the seventeenth and eighteenth centuries.

Lakatos's case study is concerned with a conjecture about polyhedra first published by Euler in 1758, though, as Lakatos points out (1963–4, I, p. 7, Footnote 1, 1976, p. 6, Footnote 1), a similar conjecture had been formulated by Descartes in a manuscript of around 1639. Euler's version is concerned with vertices (V), edges (E) and faces (F). Euler's conjecture was that, for all polyhedra, $V - E + F = 2$. Euler verified this result for all the polyhedra he could think of but did not give a general proof of the result. In 1811, however, Cauchy read a paper which gave a proof.

We see from all this that Lakatos's case study is concerned with an important piece of mathematics. The study of polyhedra was considered very significant both in ancient Greek times and in the seventeenth century because it was thought that the five regular or Platonic solids might provide clues to the nature of the cosmos. The conjecture about polyhedra was put forward by Euler, one of the leading mathematicians of the eighteenth century, and had been anticipated by Descartes. The first proof of the Descartes–Euler conjecture was given by Cauchy, one of the leading mathematicians of the nineteenth century. A case study in the history of mathematics can hardly have better credentials than this.

Lakatos begins by giving a proof of the Descartes–Euler conjecture, which is not exactly the same as Cauchy's original proof, but which is based on Cauchy's proof-idea. Lakatos then introduces refutations, reactions to these refutations and variant proofs. His discussion of the example is quite long and complicated. Here, I will give just a selection from this discussion, but one which is designed to bring out some of Lakatos' most important points.

---

[3] This passage is cited by Lakatos (1978b, p. 84, Footnote 4). The citation comes from Lakatos's PhD thesis of 1961 but was first published in his *Philosophical Papers Volume 2* of 1978.

Figure 1 Removing triangles from the plane network

I will begin with the proof that $V - E + F = 2$ for any polyhedron. Let us take an arbitrary polyhedron and suppose that $V - E + F = n$. We have to show that $n = 2$. *Step* 1. Let us imagine that the polyhedron is hollow with a surface of thin rubber. We begin by cutting out one face. This changes $V - E + F$ from n to $n - 1$. We then stretch the resulting rubber surface until it lies flat on a plane. For the resulting plane network, we still have $V - E + F = n - 1$. *Step* 2. Next, we triangulate the plane network by joining vertices of the faces until they all consist of triangles. Each step of this procedure adds one edge and one face, so that $V - E + F$ is still $n - 1$. *Step* 3. We then remove the triangles of the network one by one. There are two possible cases shown in Figure 1.[4]

Removing the triangle shown in black in 1(a), removes one face and one edge. So, $V - E + F$ remains the same, that is, $n - 1$. Removing the triangle shown in black in 1(b), removes one vertex, two edges and one face. So, $V - E + F$ remains the same, that is, $n - 1$. At the end of this procedure, we get a single triangle for which $V - E + F = n - 1$, but for a single triangle $V - E + F = 1$. Hence $n = 2$. Q.E.D.

I think anyone studying this proof for the first time would find it both very elegant and entirely convincing. However, a counterexample to the theorem was found soon after Cauchy's proof of 1811. This consisted of a pair of nested cubes as shown in Figure 2.[5]

We can imagine this obtained from a solid cube (the outer cube) by hollowing out an inner cube which does not touch the outer one. For each cube $V - E + F = 2$. So, for the hollow cube $V - E + F = 4$. We thus have a counterexample to the theorem. This is the first of Lakatos's refutations. It was discovered by Lhuilier in 1812–13 and rediscovered by Hessel in 1832. According to Lakatos (1963–4, I, p. 15, Footnote 1, 1976, p. 13, Footnote 1):

> Both Lhuilier and Hessel were led to their discovery by mineralogical collections in which they noticed some double crystals, where the inner crystal is not translucent, but the outer is. Lhuilier acknowledges the stimulus of the crystal collection of his friend Professor Pictet . . . Hessel refers to lead sulphide cubes enclosed in translucent calcium fluoride crystals.

---

[4] These are Figures 3(a) and 3(b) in the original text. The figures show the triangulated plane network for a cube.

[5] This is Figure 5 in the original text.

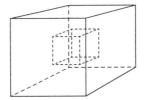

Figure 2 Pair of nested cubes

It is interesting to note that, while the attempts of both Plato and Kepler to apply polyhedra to the natural world resulted in theories which were false, the theory of polyhedra has been successfully applied in the study of crystals.

So, we now have a proof of the Descartes–Euler conjecture, and also what seems to be a refutation of this conjecture. What should be done about this situation? There seem to be two possible reactions:

(a) The pair of nested cubes is a genuine polyhedron and shows that the Descartes–Euler conjecture is false.
(b) The pair of nested cubes is not a genuine polyhedron, and the Descartes–Euler conjecture holds good.

For reaction (b), Lakatos invented the vivid name of 'monster-barring'. The idea is that, as Lakatos puts it (1963–4, I, pp. 15–16, 1976, p. 14): 'This pair of nested cubes is not a counter-example at all. It is a *monster*, a pathological case not a counter-example'. This approach was adopted historically by Jonquières. Lakatos quotes the following passage about the pair of nested cubes from a paper of Jonquières published in 1890 (1963–4, I, p. 16, Footnote 2, 1976, p. 15, Footnote 1):

> Such a system is not really a polyhedron but a pair of distinct polyhedra, each independent of the other. . . . A polyhedron, at least from the classical point of view, deserves the name only if, before all else, a point can move continuously over its entire surface; here this is not the case . . . This first exception of Lhuilier can therefore be discarded.

This issue clearly hinges on the meaning of the word 'polyhedron'. As we saw Euclid introduced the term by giving a number of examples (pyramid, prism, octahedron, icosahedron, and dodecahedron), and then the general term on the basis of these examples. The assumption here was that the introductory examples were sufficient to enable any future case to be classified. However, the example of the pair of nested cubes shows that this was not the case. Some mathematicians regarded a pair of nested cubes as a polyhedron, while others denied that it was a polyhedron, even though all these mathematicians agreed that the introductory examples (pyramid, prism, octahedron, icosahedron, and dodecahedron) were polyhedra.

One way of trying to clear up the matter is by introducing some new definitions of the word 'polyhedron'. Lakatos in fact considers two such definitions (1963–4, I, p. 16, 1976, p. 14):

> *A polyhedron is a solid whose surface consists of polygonal faces.* . . . *Def.1.*
> *A polyhedron is a surface consisting of a system of polygons.* . . . *Def.2.*

Now Def. 2 of a polyhedron as a surface favours Jonquières' view that the pair of nested cubes is not a single polyhedron but a pair of distinct polyhedra. Euclid, however, seems to adopt Def. 1, since he speaks of a polyhedral solid and of the five regular solids. Even on this definition, however, the pair of nested cubes might be ruled out if we exclude hollow solids. Euclid, however, seems to allow rather than exclude hollow polyhedra. He writes (Elements XII, Proposition 17, p. 363):

> *Given two spheres about the same centre, to inscribe in the greater sphere a polyhedral solid which does not touch the lesser sphere at its surface.*

Euclid then goes on to construct a polyhedral solid whose vertices lie on the greater sphere, but whose inner surface encloses the lesser sphere without touching it. In effect this is a hollow solid. So, it is likely that Euclid would have accepted the pair of nested cubes as a genuine polyhedron (polyhedral solid).

This discussion by Lakatos casts doubt on the view of mathematics as consisting of eternal truths. An eternal truth must presumably be expressed in some sentence S, but the meanings of the words of S may not be completely fixed. It is in fact quite typical that the meaning of a word is explained by giving a number of examples. It is assumed that these examples fix the meaning of the word with sufficient precision for it to be applied in future cases. Yet, as new cases come to light, there may well be disagreements about whether the word applies to them or not. To put it another way, concepts may change with time, and it can also happen that what was once thought as a single unified concept is split into a number of different concepts. In the present example, the unified concept of polyhedron is split into 'polyhedron qua solid', and 'polyhedron qua surface'. So, a sentence does not express a timeless truth, but a truth which may hold for the meanings of the words in the sentence at a particular time but may no longer hold as the words acquire new meanings with the development of knowledge.

Earlier we quoted Frege as saying 'the Pythagorean theorem is timelessly true'. But the Pythagorean theorem involves the concept of triangle and hence that of straight line. However, Euclid's concept of straight line was changed radically with the discovery and development of non-Euclidean geometry. In Euclidean

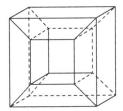

Figure 3 Picture-frame

geometry the great circles of a sphere are not straight lines. Yet in the non-Euclidean geometry of the surface of a sphere, the great circles can be straight lines. When considering any theorem about straight lines, we have to have a concept of straight line, and this concept has changed historically. This is a strong argument against the view of mathematics as consisting of timeless truths, and in favour of a historical approach to the philosophy of mathematics.

Let us now consider another of Lakatos's refutations. This is the *picture-frame* (see Figure 3[6])

We can obtain the picture-frame from the nested cubes by joining the corresponding vertices of the inner and outer cube and then 'hollowing out' from two opposite sides. For the nested cubes, $V - E + F = 4$. In the 'hollowing out' process, we add on each side 4 edges and 4 faces but subtract 2 faces. So, the change in $-E + F$ is $-8 + 8 - 4 = -4$. As V remains constant, we have $V - E + F = 0$, and the picture-frame is a counterexample to the Descartes–Euler theorem. This counterexample was also discovered by Lhuilier in 1812–13.

As with the previous counterexample, the method of monster-barring could be applied and it was indeed applied by Jonquières, whom Lakatos calls (1963–4, I, Footnote 1, p. 23, 1976, Footnote 1, p. 21) 'the indefatigable monster-barrer'. In the case of the nested cubes, the method of monster barring was not unreasonable, since the counterexample could be eliminated by switching from the concept of polyhedra as solids to the concept of polyhedra as surfaces. However, redefinitions of polyhedron to eliminate the picture-frame as a monster look much more arbitrary, and so Lakatos introduces another response to the refutation, which he describes as follows (1963–4, II, p. 130, 1976, p. 33):

> *Improving the conjecture by the method of lemma-incorporation. Proof-generated theorem versus naïve conjecture* (Italics in original)

This response involves considering the proof and Lakatos introduces the following definition of proof (1963–4, I, p. 10, 1976, p. 9):

---

[6] This is Figure 9 in the original text.

I propose to retain the time-honoured technical term 'proof' for a *thought-experiment ... which suggests a decomposition of the original conjecture into subconjectures or lemmas*, thus *embedding it* in a possibly quite distant body of knowledge. Our 'proof', for instance, has embedded the original conjecture – about crystals, or, say solids – in the theory of rubber sheets.

The proof given earlier consists of three steps. We can consider lemma 1 as the assumption that step 1 can be carried out, and so on. Thus, the proof is divided into three lemmas. Now if a counterexample, such as the picture-frame, is discovered to the theorem proved, it will also be a counterexample to one or more of the lemmas. In fact, the picture-frame is a counterexample to lemma 1 as Lakatos explains in the following passage (1963–4, II, p. 130, 1976, p. 33):

To help your imagination, I will tell you that those and only those polyhedra which you can inflate into a sphere have the property that, after a face is removed, you can stretch the remaining part onto a plane [i.e. the property that lemma 1 holds – D.G.].

It is obvious that such a 'spherical' polyhedron is stretchable onto a plane after a face has been cut out; and vice versa it is equally obvious that, if a polyhedron minus a face is stretchable onto a plane, then you can bend it into a round vase which you can then cover with the missing face, thus getting a spherical polyhedron. But our picture-frame can never be inflated into a sphere; but only into a torus.

The method of lemma-incorporation now consists in replacing the original naïve conjecture, that is, for any polyhedron, $V - E + F = 2$, by the proof-generated theorem, that is (roughly) for any polyhedron which satisfies lemma 1, $V - E + F = 2$. This deals with the picture-frame refutation.

Lakatos regards lemma-incorporation as superior to monster-barring because it improves the conjecture. As he says (1963–4, II, p. 134, 1976, p. 37):

Proofs, even though they may not *prove*, certainly do help to *improve* our conjecture. ... *Our method improves by proving.*

This is a heuristic view of proof.

Later in the same paper, Lakatos puts essentially the same point in a more dramatic fashion which perhaps contains some characteristic exaggeration. He says (1963–4, IV, pp. 320–1, 1976, pp. 89–90):

The impact of proofs and refutations on naïve concepts is much more revolutionary ... they *erase* the crucial naïve concepts completely and *replace* them by proof-generated concepts. The naïve term 'polyhedron', even after being stretched by refutationists, denoted something that was crystal-like, a solid with 'plane' faces, straight edges. The proof-ideas swallowed this naïve concept and fully digested it. In the different proof-generated theorems we have nothing of the naïve concept. That disappeared without trace. Instead each proof yields its characteristic proof-generated concepts, which refer to stretchability, pumpability, ... and the like. The old problem disappeared, new ones emerged. After Columbus one should not be surprised if *one does not solve the problem one has set out to solve.*

2.3 Lakatos' Impact and His Last Work in Philosophy of Mathematics

I have only mentioned a few of the examples and discussions to be found in Lakatos's paper 'Proofs and Refutations'. Yet they suffice to show that his application of the historical method to this case history yielded some novel and quite startling ideas in the philosophy of mathematics. Popper had produced the schema of 'conjectures and refutations' for science, and Lakatos adapted it for mathematics as 'proofs and refutations'. Yet the two schemas are very different. 'Conjectures and refutations' as applied to science is very much in accordance with common sense. Scientists devise conjectures to explain phenomena. They then test these conjectures against the empirical data. If a conjecture is refuted by the data, it has to be changed. This is just the method of trial and error and seems very reasonable. 'Proofs and refutations' is quite another matter. A proof is normally thought to establish the truth of a theorem, so that refutations shouldn't occur at all. Of course, it is recognized that mistakes can occur in proofs, but this is normally thought to reflect the incompetence of the mathematician who produced the proof. A mistake should be rectified as soon as possible, and then quietly forgotten. Lakatos's view of the matter is very different. A proof does not necessarily establish the truth of a theorem, but rather divides the original conjecture into a number of subconjectures or lemmas. Refutations of proved theorems can and do appear, and they have a positive role in developing mathematical knowledge. They improve the original naïve conjecture turning it into a better conjecture which is sometimes generated by the proof. Proofs in the standard conception are about justifying theorems, but Lakatos gives proofs a heuristic role. He also shows how the process of proofs and refutations changes the concepts used in the original conjecture. Thus, for example, the original concept of 'polyhedron' is changed in the course of trying to prove the Descartes–Euler theorem that $V - E + F = 2$, particularly in the light of counter-examples to the theorem such as the pair of nested cubes, and the picture-frame. This historical change in mathematical concepts casts doubt on the traditional view of mathematics as consisting of timeless truths.

These new and striking results given by Lakatos in his paper 'Proofs and Refutations' encouraged some other researchers in philosophy of mathematics to take up the historical approach, particularly as the three standard positions in the subject all faced considerable difficulties and seemed unlikely to produce any significant novel advances.[7] This is not to say that Lakatos's paper was considered favourably by everyone working in the field. One advantage of the method of historical case studies is that it is difficult to deny altogether features which have been shown to apply in at least one case. However, critics can still say that this case is exceptional, and the features do not apply to most of mathematics. This line of criticism was used against Lakatos, and, strange to say, he himself partly agreed with it.

Lakatos actually gave a second example of the use of the method of proofs and refutations in mathematics in his PhD thesis. This has been reprinted as Appendix 1 (pp. 127–41) of Lakatos (1976). Interestingly it also concerns a theorem proved by Cauchy. In 1821, Cauchy published a proof that the limit of any convergent series of continuous functions is continuous. However, this was refuted by series of trigonometrical functions. The problem was resolved by Seidel in 1847, who spotted the guilty hidden lemma in Cauchy's proof. He provided a new proof that used the proof-generated concept of uniform convergence. Lakatos comments on this as follows (1976, p. 127):

> The method of proofs and refutations is a very general heuristic pattern of mathematical discovery. However, it seems that it was discovered only in the 1840s and even today seems paradoxical to many people; …

Now a lot of mathematics was discovered before the 1840s. So, if what Lakatos says here is correct, there must be many other heuristic patterns of mathematical discovery besides the method of proofs and refutations.

This might seem to be an argument against the significance of Lakatos' work, but instead could be seen as a factor encouraging others to take up his approach. After all, if Lakatos succeeded in discovering one important heuristic pattern in mathematical development, but there are many more to be discovered, then other researchers who adopted his historical approach might well obtain some interesting results. Moreover, Lakatos had also provided a suggestion about how further heuristic patterns in the development of mathematics could be

---

[7] This was my own case. Between 1962 and 1966, I was an undergraduate at Cambridge studying mathematics and philosophy. During the academic year 1965–6, I was hoping to go on to do a PhD in philosophy of mathematics and read the recent literature in the subject looking for a hopeful line of research in the field. Lakatos' paper 'Proofs and Refutations' seemed to me far and away the most interesting and exciting paper in philosophy of mathematics to have been published in the previous few years. So, I wrote to Lakatos asking if he could take me on as a PhD student. He agreed to do so, and I started my PhD with him in the autumn of 1966. I was in fact his first PhD student. Later, I discovered that several other people in different countries had been similarly impressed with 'Proofs and Refutations' and had decided to adopt its approach, though without coming to LSE.

discovered, namely: consider heuristic patterns which have been discovered in the development of science, and see whether these could be applied, perhaps in a modified form, to the development of mathematics.

In this context, it is interesting to look at Lakatos' last piece of research in philosophy of mathematics. Almost all, Lakatos' published work in philosophy of mathematics was written before the end of 1966. After that date, he switched his attention from philosophy of mathematics to philosophy of science.[8] In the period from 1959 to 1966, Lakatos was a great admirer of Popper and largely followed Popper's views on philosophy of science. He began his own research in the field by criticizing Carnap's views on inductive logic from a broadly Popperian perspective. However, Lakatos then became critical of his former master, and, using the Duhem–Quine thesis, he attacked Popper's key notion of falsification. Lakatos also studied and criticized Kuhn's views in the philosophy of science. He eventually produced his own system which is called the 'Methodology of Scientific Research Programmes' (MSRP) and is supposed to steer a happy middle course between Popper and Kuhn. During this period of work in philosophy of science, Lakatos had by no means forgotten the philosophy of mathematics. His completed works in philosophy of mathematics had made considerable use of Popper's philosophy of science, much of which he had now rejected. He therefore planned to redo his work in philosophy of mathematics, using instead of Popper's views on philosophy of science, his own MSRP. In 1973, he gave a talk which was in effect the first step in this project, but, as he died on 2 February 1974, he did not make any further progress. It is worth looking briefly at this first step which, however, is rather sketchy in character.

The paper is concerned with the method of analysis-synthesis which was invented by the ancient Greeks. As we might expect, Lakatos introduces the concept of research programme into the discussion, writing (1973, p. 99):

> *Analysis is only revolutionary when it engineers a breakthrough from a low-level naïve conjecture to a research programme.* (Italics in original)

---

[8] The analysis of Lakatos' writings agrees with my own experience. When I started my PhD with him in the autumn of 1966, I was intending to work on a project in the philosophy of mathematics along the lines of 'Proofs and Refutations'. I went over with Lakatos his more recent papers in philosophy of mathematics, some of which had not yet been published and existed only as typescripts. However, Lakatos was then working on probability and inductive logic, and this led me to change my PhD topic to the philosophy of probability, which, though mathematical, is normally considered part of philosophy of science. During the time I knew Lakatos, from 1966 to his death, he worked almost exclusively on philosophy of science, but was also laying plans to return to philosophy of mathematics and develop his views in the light of his new ideas about philosophy of science.

He then goes on, with a reference to his friend Árpad Szabó, to use this point of view to explain the emergence of the Euclidean axiomatic system (1973, pp. 99–100):

> It was only after hundreds of successful analyses and syntheses, after hundreds of 'proof-procedures' (in the sense of my *Proofs and Refutations*) that certain lemmas kept cropping up, became 'corroborated' (their alternatives remaining sterile) and finally were turned into the hard core of a research programme (an 'axiomatic system') by Euclid.

A couple of comments can be made on this interesting passage. First of all, Lakatos seems here to argue that the method of proofs and refutations was used in pre-Euclidean Greek mathematics, though in a passage quoted earlier (1976, p. 127) he says that the method was invented in the 1840s. As our knowledge of pre-Euclidean Greek mathematics is rather fragmentary, this would be hard to establish. Secondly, Lakatos considers the axioms of Euclid as the hard core of a research programme. This would indeed seem to be a good way of applying his MSRP to the development of mathematics. Someone who preferred Kuhn's framework to that of Lakatos, might consider Euclid's axiomatic system as defining the paradigm of ancient Greek mathematics.

Lakatos would undoubtedly have considered this example in more detail, as well as others, had he lived, but he died before he could continue with this research. His death did not, however, bring an end to the application of the historical method to the philosophy of mathematics, as I will show in Sections 3 and 4, which deal with the legacy of Lakatos in the philosophy of mathematics (1975–2023).

3 Lakatos' Legacy in the Philosophy of Mathematics I (1975–95)

I will begin by explaining what I mean by 'Lakatos' legacy'. I have suggested in Section 1 that Lakatos' principal contribution was to introduce the historical approach to philosophy of mathematics, which had not been done before. So, I will regard Lakatos' legacy as consisting of those works in philosophy of mathematics which definitely adopted the historical approach. The authors who produced these works had all, I think, read Lakatos' writings on philosophy of mathematics and were to some extent influenced by them. This does not mean, however, that they agreed with everything that Lakatos wrote, or that they were not influenced by other people as well. There are some philosophers, who are considered to be gurus, and who attract a group of faithful disciples that studies the works of 'The Master' with great care to discover more deep insights. Lakatos was no guru. His works did not inspire devotion, but they did stimulate other researchers to produce new ideas and developments, which often were of a kind of which Lakatos himself would have disapproved. In fact, I noticed in preparing this part of the book that those researchers who were strongly influenced by some of Lakatos' ideas were usually critical (often quite harshly critical) of other ideas of his. In a way this situation is not surprising. A guru philosopher is one who produces a system which others find harmonious and satisfying. Lakatos never produced such a system because he was constantly changing and developing his ideas. It is typical of him that he accepted most of Popper's philosophy of science in the period 1956–66, but then rejected a lot of it in the period 1966–74. Lakatos always pursued novelty and innovation, and in the process threw out suggestive ideas which others took up and developed further. Thus, his legacy consists of much new and original work of the kind which is unlikely to be produced by the faithful disciples of a guru philosopher.

If we understand 'Lakatos' legacy' in this rather loose sense, it turns out to quite extensive, and there is a problem about describing it in a short space. I have resolved this problem by taking just a sample from the legacy. This consists of a selection of eleven works (either papers or books) which were published between 1975 and 2023. They are the following: Crowe (1975), Dauben (1984), Dunmore (1992), Gillies (1992b), Giorello (1992), Mancosu (1996), Corfield (2003), Kvasz (2008), Guicciardini (2009), Grosholz (2016), and Cellucci (2022). This list is subject to two limitations. Many works which have made significant contributions to Lakatos' legacy have not been included for reasons of space. Hallett (1979) and Koetsier (1991) are examples. Moreover, of those authors included, I will discuss only one of their works, the one given in the list, despite the fact that many of these researchers have written other works which could be considered as part of the legacy. Thus, my account of Lakatos' legacy is only partial and far from complete. I have listed the works selected in chronological order and will consider them in this order except that I will discuss Guicciardini (2009) immediately after Mancosu (1996). This is because these two works are related to each other and also to Giorello (1992). So much then for preliminaries, let us now start looking in more detail at the first work on my list.

3.1 Crowe (1975)

Michael Crowe published his paper: 'Ten 'laws' concerning patterns of change in the history of mathematics' in 1975, the year after Lakatos' death. He refers on the first page of the paper (Crowe, 1975, p. 15) to 'The late Imre Lakatos' *Proofs and Refutations* (1963–4)' and says that it is an example of works 'that may pave the way to a new historiography of mathematics'. He goes on to say that (p. 16):

The present paper has been written to stimulate discussion of the historiography
of mathematics by asserting ten 'laws' concerning change in mathematics, . . .

Crowe certainly succeeded in his objective because his paper stimulated a great
deal of discussion. His ten laws are all of interest, but the one which generated
the most controversy was number 10 which states that (p. 19): '*Revolutions
never occur in mathematics*' (Italics in original).

Crowe defends his Law 10 as follows (1975, p. 19):

This law depends upon at least the minimal stipulation that a necessary
characteristic of a revolution is that some previously existing entity (be it
king, constitution, or theory) must be overthrown and irrevocably discarded.

He goes on to point out that the Copernican revolution did indeed satisfy this
condition, because the previously existing theories of Ptolemy and Aristotle were
certainly 'overthrown and irrevocably discarded'. On the other hand, this condition rules out the possibility of revolutions in mathematics, since the development
of new mathematical theories does not lead to older theories being 'irrevocably
discarded'. For example, the discovery and development of non-Euclidean
geometry is sometimes claimed to be a revolution in mathematics. However,
according to Crowe's condition, it cannot be, since the discovery of non-
Euclidean geometry did not lead to Euclidean geometry being 'irrevocably
discarded'. Indeed, as Crowe himself says (1975, p. 19): 'Euclid was not deposed
by, but reigns along with, the various non-Euclidean geometries'.

These arguments of Crowe's are very persuasive, but they were challenged by
the next author to be considered.

3.2 Dauben (1984)

Joseph Dauben's paper 'Conceptual revolutions and the history of mathematics:
two studies in the growth of knowledge' was published in 1984, but he had read
an earlier version in 1978. The paper is a reply to Crowe's. Dauben agrees with
Crowe that older theories in mathematics, such as Euclidean geometry, are not
discarded in the way that has happened to some scientific theories such as
Aristotelian physics and the phlogiston theory of combustion. On the other
hand, he thinks that there have been radical innovations, which have fundamen-
tally altered mathematics, or at least one of its branches, and so are justifiably
referred to as revolutions, even though they have not led to any earlier mathematics being 'irrevocably discarded'. As he says (1984, p. 52):

Revolutions have occurred in mathematics. However, because of the special
nature of mathematics, it is not always the case that an older order is refuted
or turned out. Although it may persist, the old order nevertheless does so

under different terms . . . . Often, many of the theorems and discoveries of the older mathematics are relegated to a significantly lesser position as a result of a conceptual revolution that brings an entirely new theory or mathematical discipline to the fore.

Later on Dauben describes such a relegation as follows (1984, p. 64):

The old mathematics is no longer what it seemed to be, perhaps no longer even of much interest when compared with the new and revolutionary ideas that supplant it.

So, we might say that, on Dauben's view, in a mathematical revolution, the former mathematics is not irrevocably discarded, but suffers a considerable loss of importance. In fact, it turns out that some episodes that are generally accepted to be revolutions in science, also agree with this weaker characterisation of revolution. Consider, for example, the Einsteinian revolution. This was surely a revolution in physics, and yet, after the triumph of Einstein, Newtonian mechanics is still being taught and is still applied in a wide class of cases. On the other hand, after the success of relativistic mechanics, Newtonian mechanics has undoubtedly experienced a considerable loss of importance.

Dauben illustrates his claim that there are revolutions in mathematics by giving two detailed case studies, namely (i) the discovery of incommensurable magnitudes in Antiquity and (ii) Cantor's introduction of transfinite set theory. However, he also mentions the very first use of the term 'revolution' as applied to either science or mathematics. This was in a quotation by Bernard de Fontenelle and concerned the discovery of the calculus by Newton and Leibniz. More specifically Fontenelle, writing in 1719, is referring to the Marquis de l'Hôpital's book *Analyse des infiniments petits*, which was first published in 1696, and can be considered as the first textbook of the new calculus. Fontenelle says in Dauben's English translation (Dauben, 1984, pp. 51–2):

In those days the book of the Marquis de l'Hôpital had appeared, and almost all the mathematicians began to turn to the side of the new geometry of the infinite, until then hardly known at all. The surpassing universality of its methods, the elegant brevity of its demonstrations, the finesse and directness of the most difficult solutions, its singular and unprecedented novelty, it all embellishes the spirit and has created, in the world of geometry, an unmistakable revolution.

The discovery of calculus was a revolution which transformed mathematics, but it did not lead to the rejection of any earlier mathematics. In fact, even admirers of the calculus in the early eighteenth century often regarded Euclidean geometry as superior in rigour.

Dauben gives a similar analysis in his two detailed case studies. He says (1984, p. 52):

> It is ... possible to interpret the discovery of incommensurable magnitudes in Antiquity as the occasion for the first great transformation in mathematics, namely, its transformation from a mathematics of discrete numbers and their ratios to a new theory of proportions as presented in Book V of Euclid's *Elements*.

and (1984, p. 62):

> Cantor's proof of the non-denumerability of the real numbers ... led to the creation of the transfinite numbers. This was conceptually impossible within the bounds of traditional mathematics, yet in no way did it contradict or compromise finite mathematics. Cantor's work did not displace, but it *did* augment the capacity of previous theory in a way that was revolutionary ...

3.3 Dunmore (1992)

It is time now to describe how I became involved in the question of revolutions in mathematics. My attention was first drawn to the debate between Crowe and Dauben by Caroline Dunmore, whose PhD thesis on the philosophy of mathematics I was supervising. Caroline Dunmore had adopted the historical approach to the subject, and, while reading the literature, had come across the question of revolutions in mathematics. She thought this would be a good topic for inclusion in her thesis. She did indeed include it. Her PhD was awarded by the University of London in 1989 for a thesis with the title: 'Evolution and revolution in the development of mathematics'.

It is strange that I missed the papers by Crowe and Dauben when they first appeared. I had met Joseph Dauben in New York in 1982 when I was on a spell of sabbatical leave in the USA. I had read his 1979 book: *Georg Cantor. His Mathematics and Philosophy of the Infinite* and wanted to discuss it with the author. Conversely, I met Michael Crowe when he was on a period of sabbatical leave in England in 1986–7. In fact, he gave a talk to our departmental seminar in London in October 1986 – though not on revolutions in mathematics! Once I had got to know about the Crowe–Dauben debate, I found it most intriguing, and decided not only to try to contribute to it myself, but also to edit a collection of papers on the subject. This duly appeared as Gillies, 1992a, *Revolutions in Mathematics*. The papers by the original participants in the debate (Crowe, Dauben, and Mehrtens) were reprinted, and their authors added an appendix or afterword giving their further thoughts on the subject. In addition, there were papers by a further nine authors, covering a wide range of historical cases which might be considered as revolutions in mathematics.

One thing that is striking about this collection is its international character. Of the twelve authors, three were American, three were British, two were German, one (Yuxin Zheng) was from Nanjing in Mainland China, and three were Italian. However, of the three Italians, only one (Giulio Giorello) was still in Italy, Paolo Mancosu was in the USA at Yale, while Luciano Boi was in Germany at the time, but went on to have a career in Paris. Nowadays, mathematics plays an essential role in the economies of industrialized countries. So, it is perhaps not surprising that the philosophy of mathematics has a wide international appeal. It is also interesting to note that seven of the twelve contributors argued in favour of the existence of revolutions in mathematics, three were definitely against, while two were a bit equivocal. Thus, the majority did support the existence of revolutions in mathematics, but there were several voices on the other side, so that the issue can still be considered as not entirely decided.

Dunmore (1992) 'Meta-revolutions in mathematics' is one of the papers in the volume. It was extracted from her PhD thesis. In it she gives a novel, interesting, and ingenious development of the Crowe–Dauben debate. Her starting point is the following passage from Crowe's paper (1975, p. 19):

> The stress in Law 10 on the preposition 'in' is crucial, for, as a number of the earlier laws make clear, revolutions may occur in mathematical nomenclature, symbolism, metamathematics (e.g. the metaphysics of mathematics), methodology (e.g. standards of rigour), and perhaps even in the historiography of mathematics.

Dunmore picks up the point here about metamathematics, and suggests that, while revolutions do not occur in mathematics at the object level, they do occur at the meta-level. This is how she puts it (Dunmore, 1992, p. 212):

> This is my thesis on revolutions in mathematics: revolutions do occur in mathematics, but they are confined entirely to the metamathematical component of the mathematical world. A necessary condition for a revolution to have taken place is that something formerly accepted by the community is discarded and replaced by something else incompatible with it. But what is discarded and replaced in a mathematical revolution is a metamathematical value and not an actual mathematical result.

Dunmore illustrates her account with Crowe's example of the discovery of non-Euclidean geometry. Before the discovery of non-Euclidean geometry, virtually all mathematicians held the meta-level doctrine that there was only one possible geometry, namely Euclidean geometry, that the truth of this geometry could be established *a priori*, and that this geometry was the correct geometry of space. After the discovery of non-Euclidean geometry, these meta-level doctrines were 'overthrown and irrevocably discarded' to be replaced by the view that a number of different geometries were possible. Because of these changes at the meta-level, the discovery of non-Euclidean geometry is for Dunmore a revolution in mathematics.

It is interesting to note that Dunmore in her paper also applies her account to the two principal examples given by Dauben in his paper. Regarding the discovery of incommensurables, she writes as follows (1992, p. 215):

> This is the story of perhaps the first great meta-level revolution in the development of mathematics. Pythagorean mathematics was based firmly on the cherished belief that the positive integers were all the numbers that there were, and that all phenomena could be expressed in terms of ratios of integers. But their great interest in such figures as the square and regular pentagon and their diagonals led the Pythagoreans to the discovery of pairs of incommensurable line segments. This required the abandonment and replacement of a major element of Pythagorean philosophy . . . the resulting advance was conservative at the object-level, but demanded a meta-level revolution.

As regards Cantor and transfinite numbers, Dauben has this to say (1984, pp. 59–60):

> Cantor's introduction of the actual infinite in the form of transfinite numbers was a radical departure from traditional mathematical practice, even dogma. This was especially true because mathematicians, philosophers, and theologians in general had repudiated the concept since the time of Aristotle.

Dunmore picks up this point about the actual infinite and argues that it shows that the case of Cantor and transfinite numbers fits her account of revolutions at the meta-level. She says (1992, p. 221):

> Another example of a meta-level revolution in mathematics is Cantor's invention of the theory of transfinite sets. Cantor's work on the continuum and actually infinite sets demanded . . . the abandonment of mathematicians' and philosophers' beliefs about the concept of infinity, so crucial to mathematics.

What is striking here is that Dunmore agrees with Dauben that the two case studies he presents are, in some sense, revolutions in mathematics, but the reasons she gives for this conclusion are quite different from Dauben's.

3.4 Gillies (1992b)

I will next say a few things about my own paper on this subject, but, before doing so, I would like to answer an objection that may have occurred to some readers. I have been talking about a debate on whether revolutions in mathematics occur as part of Lakatos' legacy. 'But', someone might ask, 'is this debate not part of Kuhn's legacy rather than Lakatos' legacy?' My answer is that it should be considered as part of both Kuhn's legacy and Lakatos' legacy.

Kuhn is famous for his analysis of revolutions in science, and thus considering the question of whether revolutions occur in mathematics as well as science must be considered as part of Kuhn's legacy. On the other hand, Kuhn never wrote about mathematics, only about science. Lakatos was the first to apply the historical method to the philosophy of mathematics, and he used the heuristic of considering whether a pattern of development in science applied also to mathematics – though perhaps in a modified form. This is what led him from Popper's 'Conjectures and Refutations' to his own 'Proofs and Refutations'. Now considering whether Kuhn's 'Revolutions in Science' occur also in mathematics follows the same pattern and should for this reason be considered as part of Lakatos' legacy as well as Kuhn's.

Having said this, however, it should be added that Lakatos would probably not have approved of this part of his legacy. In his researches in philosophy of science, Lakatos was highly critical of Kuhn. Here are a couple of representative quotations from Lakatos (1970):

> If even in science there is no other way of judging a theory but by assessing the number, faith and vocal energy of its supporters ... truth lies in power. (pp. 9–10)

> ... *in Kuhn's view scientific revolution is irrational, a matter for mob psychology.* (p. 91. Italics in original)

Lakatos is here objecting to Kuhn's bringing social factors into his philosophy of science, as is shown by Kuhn's frequent use of the phrase the 'scientific community'.

Lakatos, by contrast, explicitly excluded social analysis from philosophy as the following passage from his PhD thesis of 1961 shows (1976, pp. 145–6):

> The Hegelian conception of heuristic ... is roughly this. Mathematical activity is human activity. Certain aspects of this activity – as of any human activity – can be studied by psychology, others by history. Heuristic is not primarily interested in these aspects. But mathematical activity produces mathematics. Mathematics, this product of human activity, 'alienates itself' from the human activity which has been producing it. It becomes a living, growing organism, that *acquires a certain autonomy* from the activity which has produced it; it develops its own autonomous laws of growth, its own dialectic. ... The activity of human mathematicians, as it appears in history, is only a fumbling realisation of the wonderful dialectic of mathematical ideas. ... Now heuristic is concerned with the autonomous dialectic of mathematics and not with its history, though it can study its subject only through the study of history and through the rational reconstruction of history.

This quotation is of interest from various points of view. It shows that Lakatos was still strongly influenced by Hegel at the time he wrote his PhD thesis. Indeed, he says in the thesis: 'The three major – apparently quite incompatible – "ideological" sources of the thesis are Pólya's mathematical heuristic, Hegel's dialectic and Popper's critical philosophy' (Quoted from Lakatos, 1978b, p. 70, Footnote 2). This is perhaps not so surprising as it may sound at first. Lakatos came from Hungary where the dominant philosophical school was a Hegelian version of Marxism, but when he arrived in England, he became friendly with Popper, whose philosophy he studied closely. The Hegelian side of Lakatos has been emphasized by Larvor in his 1998, where he says (p. 9): 'there was always a dialectical-Hegelian element to Lakatos' work'. Larvor adds (p. 9): 'dialectical logic studies the development of *concepts*'. This characterization of dialectics applies well to Lakatos' writings.

The Hegelian quotation from Lakatos is also very relevant to Lakatos' concept of 'rationally reconstructed history' mentioned earlier in Section 2.2. This quotation shows that Lakatos' aim was to give an account of the wonderful autonomous dialectic of mathematics. To do so, he had to begin by studying the actual history of 'fumbling' human mathematicians, but this history had to be 'rationally reconstructed' to get closer to the dialectic. The rational reconstruction of history is supposed to be contained in the dialogue, but there is something problematic about Lakatos' use of the dialogue form here. Is this not introducing the social community of mathematicians discussing a problem, that is, something which Lakatos' condemned Kuhn for doing? Some commentators on Lakatos have drawn attention to this problem.

Larvor writes in his 1998, p. 23:

> Lakatos' use of the dialogue form may invite a comparison with Plato and indeed *Proofs and Refutations* shows flashes of wit and irony similar to those in Socratic dialogues. ... In a dialogue, the dialectical development of concepts is leavened by human drama between the characters. Hegel thought that the use of human mouthpieces for philosophical ideas introduces irrelevant questions which can only distract attention from the logical development of the ideas themselves. Nevertheless, there is a sense in which *Proofs and Refutations* is closer to Hegel than to Plato.

This question is also discussed by Dutilh Novaes in her interesting 2021 book, for which she won the 2022 Lakatos Award.[9] The main hypothesis of the book is that (Dutilh Novaes, 2021, p. 29) 'deductive reasoning is essentially a dialogical phenomenon'. She goes on to say (2021, p. 46):

---

[9] The Lakatos award is given annually for an outstanding contribution to the philosophy of science, widely interpreted, published in English during the previous six years. It was endowed by the Latsis foundation.

Among other proposals in the literature, the one that comes closest in spirit to the present proposal is Imre Lakatos' account of mathematical practice in terms of the dialectic of *Proofs and Refutations* . . . .

and adds the following footnote to this passage (2021, p. 46, Footnote 9):

Lakatos' work did not provide initial inspiration for my dialogical account of deduction, which was in fact mainly inspired by the historical emergence of deduction in Ancient Greece (both in logic and in mathematics). But there are some important similarities . . . and indeed engagement with Lakatos' ideas while working on the 'The Roots of Deduction' project helped me clarify a number of aspects of my proposal.

However, Dutilh Novaes goes on to point out that her position nonetheless differs in some important respects from that of Lakatos. She agrees with Larvor that Lakatos, despite his use of the dialogue form, is more concerned to uncover an underlying Hegelian dialectic than to analyse the linguistic interactions between researchers in mathematics. She writes (2021, p. 48):

Interestingly, and despite the fact that *Proofs and Refutations* is written in dialogue form, Lakatos' dialectical philosophy of mathematics is not particularly *dialogical*. . . . One might speculate that this feature is somehow related to a Hegelian disregard for specific agents, given that the objective *Geist* should transcend the individual's subjective mind. . . . if the dialogical, multi-agent component is indeed not fundamental to Lakatos' overall account after all, then this is a significant dissimilarity between the Lakatosian picture and the present proposal, in which agents and their interactions play a fundamental role.

There is a considerable irony here in that Lakatos, despite his wish to eliminate social factors from philosophy, nonetheless helped, through his work, in the development of a social account of logic based on agents and their interactions.

Lakatos believed that Kuhn, by concentrating on the scientific community and its operation, would for that reason fail to achieve the admirable goal of uncovering the wonderful dialectic of scientific ideas. Actually, Lakatos at the time of his criticisms of Kuhn had moved away from his 1961 Hegelian position; but this did not make a great deal of difference on this point. In his later writings, Lakatos replaced the Hegelian autonomous dialectic with Popper's theory of a third world (or world 3) of objective ideas. Yet, as Popper himself says (1972, p. 106):

What I call 'the third world' has admittedly much in common . . . with Hegel's objective spirit, though my theory differs radically, in some decisive respects, from . . . Hegel's.

The key point remained for Lakatos that philosophy should deal with propositions, theories, problems; and *not* with the social group producing, discussing and modifying these propositions, theories and problems.[10]

My own position diverges from that of Lakatos without however rejecting Lakatos' ideas completely. I don't think that philosophy should confine itself exclusively to the realm of objective ideas, and I do think that it is quite legitimate, as Kuhn suggests, for philosophy also to consider the social group that produces these objective ideas. Moreover, I argue in my 1992b that Kuhn's concept of paradigm differs from Lakatos' concept of research programme, but that both are needed in the analysis of the development of science and mathematics. This is my account of the difference (1992b, p. 284):

> A paradigm consists of the assumptions shared by all those working in a given branch of science at a particular time. A historian can reconstruct the paradigm of a specific group at a particular time by studying the textbooks used to instruct those wishing to become experts in the field in question. Thus a paradigm is what is common to a whole community of experts in a particular field at a particular time. By contrast, only a few of these experts (or, in the limit, only one) may be working on a particular research programme. Characteristically, only a handful of vanguard researchers will have been working on a specific research programme at a particular time. A historian who wishes to reconstruct a research programme will look not at textbooks in wide circulation, but at the writings of a few key figures. He or she will examine the notebooks, the correspondence, and the research publications of these leading figures, and go on to reconstruct the programme on which they were working. In general, then, we can say that research programmes differ from paradigms.

My 1992b paper is entitled 'The Fregean revolution in logic'. I argue that it is similar to the Copernican revolution in astronomy in that Frege, like Copernicus, began the revolution, but did not complete it. The revolution began with Frege's publication of his 1879 *Begriffsschrift* but was not completed until the early 1930s. In the analysis of this revolution, I use both the concept of paradigm and that of research programme. The relevant paradigms are described as follows (1992b, p. 266):

> Before the revolution, logic was dominated by the Aristotelian paradigm, whose core was the theory of the syllogism; after the revolution, logic came to be dominated by the Fregean paradigm, whose core was propositional calculus, and first-order predicate calculus.

These claims are confirmed by examining a pre-revolution textbook of logic of 1884 (1992b, pp. 271–2), and two post-revolution textbooks of logic of 1964 and 1977 (1992b, pp. 272–6).

---

[10] For further discussion of this issue, see Gillies (2014).

Turning now to the concept of research programme, I argue that it was Frege's work on a specific research programme, the Logicist programme, which gave rise to his advances in logic (1992b, p. 287):

> The aim of Frege's research programme (the Logicist programme) was to show that arithmetic could be reduced to logic. The programme failed achieve its goal. Frege's own attempt was vitiated by Russell's paradox, and Russell's later efforts by Gödel's first incompleteness theorem, and other difficulties. However, few successful research programmes can have been as intellectually fruitful as the unsuccessful Logicist programme. . . . the Logicist programme gave rise to a revolution in formal logic; but that is far from the end of the matter. Frege's work on his Logicist programme produced advances in philosophy and the theory of language as well . . .

My 1992b also compares and contrasts Frege's research programme with those of Boole and Peano.

I am not the only researcher who has tried to combine Kuhn's ideas with those of Lakatos. Dunmore (1992, p. 224) speaks of 'the implicit metamathematical views of the community that generate and guide their research programmes'. Giorello also combines ideas from Kuhn and Lakatos, as we shall see when his paper is analysed in the next section.

3.5 Giorello (1992)

Giulio Giorello, a professor in Milan, was an enthusiast for the historical approach to philosophy of science and mathematics, and a great admirer of Lakatos. Giorello took a particular interest in the discovery and early development of the calculus. This is obviously an episode of the very greatest importance in the history of mathematics. Yet it was hardly mentioned by Lakatos, so that Giorello's interest in the topic represents a new departure. The title of his 1992 paper: 'The "fine structure" of mathematical revolutions: metaphysics, legitimacy, and rigour. The case of the calculus from Newton to Berkeley and Maclaurin' shows that he was an admirer of Kuhn as well as Lakatos. Indeed, he speaks of (1992, p. 140) 'the Calculus revolution initiated by Newton in Britain and by Leibniz on the Continent'. However, Giorello introduces some new concepts which are not to be found in Kuhn, and which Giorello characterizes as 'the fine structure' of mathematical revolutions.

The first of these concepts is that of a 'paradigm of legitimacy'. This concept has a curious history, since it was first introduced, as Giorello points out, by René Thom in order to give a mathematical account of political revolutions. Then Giorello applied it to mathematical revolutions. Indeed, he claimed (1992, p. 165):

> A revolution in mathematics demands as a necessary (but not sufficient!) condition a violation of the previously accepted paradigm of legitimacy.

Whether this applies to all revolutions in mathematics might be questioned, but it certainly does apply in the case of the introduction of the infinitesimal calculus. Here, the old paradigm of legitimacy was the geometrical rigour of the ancients as represented by Euclid and Archimedes. Now the new calculus was a more powerful mathematical technique than ancient Greek geometry, but it did appear to lack 'the rigour of the ancients'. Giorello illustrates this with the simple example of finding the tangent to the parabola $y = x^2$. If we carry out the calculation using the Leibnizian infinitesimals dy, dx, we find moves which are distinctly suspicious. At one point, we divide the equation through by dx to get dy/dx. This would only seem justified if dx is non-zero. However, further on dx is set equal to zero to obtain the result.

Leibniz attempted to validate such moves partly by an appeal to a metaphysical principle (the law of continuity), and partly from the fact that the results of the new calculus agreed with those of ancient Greek geometry when the latter were obtainable. As Giorello puts it (1992, p. 150):

> Infinitely small and infinitely large numbers are well-founded fictions (*fictions bien fondées* or *fondées en realité*) because (*this is the deductive argument*) their use follows from the law of continuity, or because (and this is the inductive argument) wherever results obtained by 'the method of the Ancients' are available, the new geometry and the old are seen to coincide.

This was Leibniz' approach, but Newton's approach was different, and Giorello also describes a third approach due to the Dutch philosopher and mathematician Bernhard Neuwentijdt (Giorello, 1992, pp. 146–8). Giorello then argues that we can clarify the situation by introducing Lakatos' concept of research programme. As he says (1992, p. 160):

> Having *de facto* violated the old standards of rigour, and lacking a new paradigm of legitimacy, the innovators had to fall back on some metaphysical principles (the law of continuity for Leibniz, the intuition of physical motion for Newton). These metaphysical principles were articulated into distinct 'research programmes' ...; these research programmes in mathematics are in many ways comparable to Lakatos's research programmes in the empirical sciences. ... Whereas there was agreement as to the need for restoring the strictness of the Ancients, there was still disagreement as to the particular recipe to be adopted, or, as eighteenth-century mathematicians used to say, as to the particular kind of 'metaphysics' by which to rescue the procedures of the calculus.

Giulio Giorello gave lectures on these topics in Milan, and he fired an interest in them in two students who attended these lectures, namely Paolo Mancosu and Niccolò Guicciardini. They went on to make some novel and original contributions in these, and related, areas, as we shall see in the next section.

4 Lakatos' Legacy in the Philosophy of Mathematics II (1996–2023)

4.1 Mancosu (1996)

The title of Mancosu's 1996 book *Philosophy of Mathematics & Mathematical Practice in the Seventeenth Century* gives a good idea of some of the innovations it contains. First of all, it promises an account of philosophy of mathematics in the seventeenth century, something which had never been given before. Nearly all previous expositions of philosophy of mathematics had, after covering the ancient Greek period, restarted the story with Kant. Trying to reconstruct the philosophy of mathematics of the seventeenth century can be considered as carrying the historicization of philosophy of mathematics a stage further than Lakatos. Lakatos had introduced the historical approach to philosophy of mathematics, but should the historical approach not be applied to philosophy of mathematics itself, and an investigation carried out as to how philosophical ideas about mathematics had changed historically? This was a step which Lakatos himself did not take.

Secondly, the book considers the relation between philosophy of mathematics and mathematical practice. Mancosu may have here been influenced by Giorello, who, as we saw in the previous section, applied Lakatos' concept of research programme to the development of mathematics and argued that, in the late seventeenth and early eighteenth centuries, research programmes were introduced which aimed to produce a more rigorous development of the calculus. These research programmes were guided by metaphysical principles, and this suggest an influence of philosophy on mathematical practice. Mancosu explores this influence in the seventeenth century and also shows that, conversely, new mathematical results led to changes in philosophy.

Interestingly, this theme of philosophy interacting with mathematical practice applies just as much to the three classic philosophies of mathematics of the late nineteenth and twentieth century – logicism, formalism, and intuitionism. Logicism guided a mathematical research programme which led to the development of contemporary mathematical logic. Formalism guided Hilbert's programme which led to the development of metamathematics and Gödel's incompleteness theorems. Intuitionism guided Brouwer in his development of a new constructive mathematics, and related philosophical ideas led to other versions of constructivism. The philosophical theories which guided all three schools encountered severe difficulties. Gödel's incompleteness theorems reduced the plausibility of both logicism and formalism, thereby giving a good example of how new mathematical results can influence philosophy. Mainstream mathematicians did not, as Brouwer had hoped, abandon their usual mathematical procedures for the new constructive ones. Yet, despite these problems, the programmes gave rise to new and useful branches of mathematics. Mathematical logic and meta-mathematics have proved very helpful in the development of computing, while constructive methods also have applications in specific areas. Of course, this account of these philosophies of mathematics is very different from the intentions of those who introduced them. They had a non-historical view of philosophy of mathematics which they saw as aiming to give a general philosophical account of the timeless truths of mathematics. Instead, we can regard their philosophies as heuristics for research programmes in the contemporary mathematical scene – research programmes which actually proved to be quite fruitful.

Returning to the seventeenth century, Mancosu has an interesting chapter (6) on the foundations of Leibniz' calculus. He recounts a foundational debate on this in the Paris Academy of Sciences from 1700 to 1706. Leibniz was invited to express his opinion by the supporters of his calculus, including L'Hôpital, but Leibniz did not seem to them to have supported it strongly enough. As Mancosu says (1996, p. 172):

> Leibniz had not expressed any commitment to infinitesimal quantities and L'Hôpital got to the point of asking Leibniz not to write anything more about the matter.

Apart from the dramas of the calculus, there were many other, less well-known but still very interesting issues in the philosophy of mathematics of the seventeenth century, and Mancosu's book gives a good account of these. Mancosu argues for the importance of Aristotle in the philosophy of mathematics of the seventeenth century, claiming that (1996, p. 102): 'a great part of the philosophy of mathematics in the seventeenth century was shaped by the Aristotelian notion of science'. Aristotelian ideas were in fact used to criticize mathematics. Aristotle had distinguished between demonstrations which simply showed that a fact was so (*hoti*) and those which showed the cause of its being so (*dioti*). Here cause does not mean just efficient cause but can be any of Aristotle's four causes, so that we might characterize the second type of demonstration as one which provides an explanation of the fact. The critics of mathematics claimed that mathematical proofs were demonstrations only in the first or weaker sense, so that mathematics failed to be a proper Aristotelian science.

As Mancosu shows, mathematicians of the seventeenth century took this criticism seriously and responded to it in a variety of ways. Some argued that all mathematical proofs were causal in Aristotle's sense, while others thought that only some proofs were causal. In particular it was widely held that proofs by contradiction were not explanatory. Indeed, Aristotle himself had explicitly asserted in *Posterior Analytics I.26* that direct proofs are superior to proofs by contradiction. These considerations led some seventeenth century mathematicians to try to develop mathematics without using proofs by contradiction. An example is Guldin in his *Centrobaryca* (1635–41). As Mancosu says (1996, pp. 27–8):

> Guldin ... set out to show that proofs by contradiction could be eliminated
> from the development of Euclidian and Archimedean mathematics.

Here is another convincing example of philosophical ideas influencing mathematical practice.

However, Mancosu also gives a good example of new mathematical results influencing philosophical ideas. Torricelli in 1641 discovered an example of a solid which is infinitely long, but yet has a finite volume. This is obtained by taking a branch of the hyperbola $xy = a^2$, say in the first quadrant, rotating it round the y-axis, and then considering the solid lying above a fixed plane $y = c$, for a constant $c > 0$. Torricelli proved that the infinitely long solid thus obtained had a finite volume.

Torricelli was virtually unknown when he became Galileo's successor as professor of mathematics at Florence in 1642, but two years later he had become one of the most famous mathematicians in Europe as the result of communicating his discovery to the French geometers in 1643 and publishing it in 1644.

Torricelli's discovery gave rise to a whole series of interesting philosophical discussions. Part of the problem arose once again from Aristotelian philosophy, since Aristotle had stated (*De Caelo I.6*) that there is no proportion between the infinite and the finite. Yet Torricelli's result seemed to show the opposite.

The discovery also posed problems for the negative theory of infinity. This was based on the idea that our talk of infinity is nothing but talk about our inability to conceive limits. Yet Torricelli's result seemed to provide positive knowledge about an infinitely long sold.

Though Torricelli's infinitely long solid with finite volume is not much discussed today, it could be argued that it still poses problems for the philosophy of mathematics. The existence of such a solid can easily be proved within Euclidean geometry; but how is this result to be interpreted and what does it show about the nature of the infinite?

4.2 Guicciardini (2009)

In the preface to his 2009 book: *Isaac Newton on Mathematical Method and Certainty*, Niccolò Guicciardini says that he will (p. xiii) 'focus on one aspect of Newton's mathematical work that has so far been overlooked, namely, what one could call Newton's philosophy of mathematics'. This plan for reconstructing the philosophy of mathematics of a seventeenth century thinker is, of course, highly reminiscent of Mancosu (1996), but Guicciardini points out that there is a gap in Mancosu's book (2009, p. xiv, Footnote 3):

> In Mancosu, *Philosophy of Mathematics and Mathematical Practice in the Seventeenth Century* (1996), an informed and thorough study of the philosophy of mathematics in the seventeenth century, Newton's name occurs only three times and in passing references.

Guicciardini's book fills this gap.

Guicciardini also mentions in the preface what is perhaps the key to understanding Newton's philosophy of mathematics (2009, p. xiii):

> Newton aimed at injecting certainty into natural philosophy by deploying mathematical reasoning.

Guicciardini goes on to give a quotation about this from Newton himself. It comes from the third of Newton's first set of lectures on optics at Cambridge, delivered between 1670 and 1672 (Guicciardini, 2009, pp. 19–20):

> I hope to show – as it were, by example – how valuable is mathematics in natural philosophy. I therefore urge geometers to investigate nature more rigorously, and those devoted to natural science to learn geometry first. Hence the former shall not entirely spend their time in speculations of no value to human life, nor shall the latter, while working assiduously with an absurd method, perpetually fail to reach their goal. But truly with the help of philosophical geometers and geometrical philosophers, instead of the conjectures and probabilities that are being blazoned about everywhere, we shall finally achieve a science of nature supported by the highest evidence.

Newton started his mathematical investigations by developing the ideas of Descartes. To Descartes' algebra, he added a new algorithm involving infinite series and fluxions, what we would now call 'calculus'. However, this powerful new mathematical tool at once raised a problem, since the reasoning that it involved appeared to be more dubious than the rigorous geometry of the ancients. Newton wanted mathematics to be certain so that it could inject certainty into natural philosophy. So, what should be done about this situation? Newton found a solution which was based on another idea of Descartes, taken in turn from Pappus, namely the distinction between analysis and synthesis. According to Descartes, his algebra belonged to 'analysis', the phase of the investigation in which the solution of the problem was discovered. This was followed by synthesis, the phase in which the correctness of the solution was rigorously proved. Newton decided that his new analysis (the calculus) should be used only as heuristic, and results discovered by its means should then in the synthesis be proved rigorously by geometry. As Guicciardini puts it (2009, p. 254):

> Newton affirmed that in writing the *Principia* he had chosen, following the practice of the ancients, to demonstrate synthetically its propositions in order to convey mathematical certainty to natural philosophy. He also claimed that most of the propositions had been first found by help of the new analysis ...

Newton's geometrical approach lost out to the more algebraic approach of Leibniz and his followers. In view of this defeat, it is easy to dismiss Newton's view as just reactionary. Guicciardini argues against such an opinion by showing that there were many arguments in favour of Newton's approach, and that it was supported by several leading mathematicians of the time.

Newton thought that curves should be considered as generated mechanically as in the simple case of a circle drawn by a compass. As Guicciardini says (2009, p. 72):

> For him, all curves, even the circle, were primarily given by mechanical descriptions. ... Curves that are posited mechanically are better understood because one knows the 'reason for their genesis'.

Newton disliked the definition of conics as sections of a cone because this was not a good way of generating them, and he preferred the use of ellipses to parabolas because they were easier to draw (Guicciardini, 2009, p. 73). Moreover, Newton may have been in contact with instrument makers and used real curve-tracing devices in his investigation of curves (Guicciardini, 2009, pp. 96–7 and p. 129). He also thought that the genesis of curves takes place 'in the reality of physical nature' through 'the motion of bodies' (Guicciardini, 2009, p. 226). This all shows Newton's very practical and applied attitude to mathematics which also comes out in the quotation from his third lecture on optics given earlier. Guicciardini sums up the situation as follows (2009, p. 302):

> Newton's downgrading of Cartesian algebra and the Leibnizian calculus to mere heuristic tools devoid of scientific character is thus based on his adoption of the ideas that the symbolism of algebra and calculus do not capture the reasons for the genesis of figures ...

Guicciardini also mentions some leading mathematicians of the time who adopted Newton's approach (2009, p. 81):

> Reading Newton's defense of geometry as a backward move and identifying algebraization as a progressive element in seventeenth-century mathematics does not capture the values that underlay the confrontation between mathematicians such as Huygens, Barrow, and Newton on the one hand and Descartes, Wallis, and Leibniz on the other.

This agrees well with Giorello's picture of two different mathematical research programmes guided by different philosophies. Guicciardini has reconstructed the philosophy which underlay Newton's approach, and also shown that it was a reasonable view for the time. He does not, however, deny that Newton's geometrical approach lost out, not only sociologically but mathematically as well. In Chapter 12 (pp. 267–90), Guicciardini gives in detail two examples from the *Principia* in which Newton could not prove his required results geometrically and had to resort to 'hidden new analysis', that is, the use of infinite series and fluxions. Guicciardini then concludes (pp. 289–90):

> Many geometrical constructions of the *Principia* only hide the fluxional analysis – most notably Newton's analytical quadrature techniques – without replacing it with a geometrical demonstration that can stand on its own independently of the analysis. ... The fact is that Newton's new analysis was far more powerful than the geometry of the ancients that he praised so consistently. Some of the results of the *Principia* could be obtained only thanks to the highly symbolical manipulations displayed in his two catalogues of curves of *De Methodis* and the new analysis of the method of series and fluxions, the innovative algorithm Newton devised in his *anni mirabiles*.

It should be remembered, however, that Newton's use of geometrical methods did make the *Principia* more convincing at the time, and so helped to establish the mechanics which it expounded. Newton's mechanics included the law of gravity involving action at a distance, and hence contradicted the standard Cartesian view of the time that action could only be by contact. If Newton had developed a dubious new mechanics using a dubious new mathematics, he might indeed have failed to convince his audience. His attempt to use only geometry was well-motivated, even though it failed. Ironically, part of the reason for Newton's failure to convince his contemporaries was Newton's outstanding brilliance as a mathematician. His use of geometrical methods was too difficult for most of his contemporaries even to understand, let alone emulate. At the same time, the results could be proved much more easily using the Leibnizian calculus. Another irony occurs because Newton's partial successes with his geometrical programme actually gave a reason for abandoning it. The fact that he was able to get as far as he did using geometry gave assurance that the new mathematics, which produced the same results more easily, was in fact correct. Newton has not been alone among the great mathematicians in pursuing a mathematical research programme that failed. After all, Hilbert's programme also failed to achieve its goal, though it did produce some results of very great significance.

Guicciardini's book won the 2011 Fernando Gil International Prize for the Philosophy of Science. This prize is a joint initiative of the Portuguese Government's Foundation for Science and Technology and the Calouste Gulbenkian Foundation, in honour of the Portuguese philosopher Fernando Gil (1937–2006). The prize is to reward a recently published work of particular excellence in the domain of philosophy of science, where science is taken in a broad sense to include mathematics. The first Fernando Gil prize was awarded in 2010.

4.3 Corfield (2003)

David Corfield has strong links with the Lakatosian tradition in philosophy of mathematics. He says (2003, p. ix): 'Imre Lakatos's *Proofs and Refutations* ... was thrust into my hands by a good friend Darian Leader, who happens to be the godson of Lakatos'. Corfield's PhD thesis, which was awarded by King's College London in 1995, was entitled: *Research Programmes, Analogy, and Logic: Three Aspects of Mathematics and its Development*. His supervisor had been a PhD student of Lakatos, so that David Corfield can be considered an academic grandson of Lakatos. His 2003 book was developed from his PhD thesis, but its title: *Towards a Philosophy of Real Mathematics* give a good indication of a new point of view which Corfield introduced into the Lakatosian tradition.

In the introduction to the book, Corfield complains that the mathematics considered by philosophers of mathematics tends to be almost exclusively the foundational mathematics of the period 1880–1930. He considers this to be (2003, p. 5) 'an unbalanced interest in the "foundational" ideas of the 1880–1930 period' and calls this attitude 'the foundationalist filter'. Corfield thinks that philosophers of mathematics should remove this filter and consider the non-foundationalist mathematics of the last seventy years which he thinks they have hitherto largely ignored. As he says (2003, pp. 7–8):

> Straight away, from simple inductive considerations, it should strike us as implausible that mathematicians dealing with number, function and space have produced nothing of philosophical significance in the past seventy years in view of their record over the previous three centuries.

The non-foundationalist mathematics of the last seventy years is the 'Real Mathematics' whose philosophy Corfield wants to explore in his book.

Corfield's interest in contemporary mathematics does not lead him to abandon the historical approach. After all, one can write the history of the last seventy years. The first topic he considers is the influence of computers on mathematics – something which indeed appeared for the first time in the period 1933–2003. He begins his treatment of this topic with the following passage praising Lakatos (2003, p. 35):

> Theorem proving, conjecturing and concept formation make up the three principal components of mathematical research. The brilliant observation of Lakatos ... was that these components are thoroughly interwoven. He was right.

Lakatos, however, was speaking of human mathematicians. What about 'artificial mathematicians'? Here Corfield observes (2003, p. 35):

> For computers to become valuable to mathematicians it will be necessary to isolate the components, or perhaps even parts of the components, mentioned above.

The first of these components is 'theorem proving' and Corfield begins by considering automated theorem proving. He makes the following insightful observation (2003, p. 38):

> I believe it is no accident that the most successful approach to date has been one that has deliberately avoided closely imitating human problem solving techniques. Computers have their own inhuman strengths which need to be harnessed.

Corfield has in mind here Larry Wos's approach to automated theorem proving, which involves translating the mathematics into the clausal form of logic – a form which is particularly suited to computers, and using methods of inference and search which are suited to computers rather than humans. A notable success for this approach occurred on 10 December 1996 when it produced a proof for a conjecture which human mathematicians had failed to prove for more than sixty years.

In 1933, Huntingdon presented a simple set of three axioms for Boolean algebra. The first two axioms were commutativity and associativity, while the third is known as the Huntingdon equation. Shortly after, another mathematician (Robbins) speculated that the Huntingdon equation could be replaced by a weaker one, known as the Robbins equation. A system with the Robbins equation replacing the Huntingdon equation became known as a Robbins algebra. Corfield says (2003, p. 48):

> It is easy to show that a Boolean algebra satisfies the Robbins equation, but it
> remained an open problem for around sixty years to determine whether
> a Robbins algebra ... is Boolean, and this despite the best efforts of Tarski
> and his students.

In 1996 a computer proof that a Robbins algebra is Boolean was produced.

This was certainly an impressive achievement, but Corfield points out that there is a problem with computer proofs. As he says (2003, p. 56):

> What mathematicians are largely looking for from each other's proofs are
> new concepts, techniques and interpretations. Computer proofs certainly give
> information concerning the truth of a result, but very little beyond this.

The problem is that, although each step in a computer proof can easily be checked and shown to be valid, the proof can still remain incomprehensible to a human. The proofs of a human mathematician are always based on an idea which is first grasped before the details are filled in. There may be no obvious idea behind a computer proof.

Corfield does however indicate a possible way out of this difficulty which consists in translating the computer proof into a language which is more accessible to humans. This approach was taken for the computer proof of the Robbins conjecture by Louis Kauffman. He translated the proof into a special symbolic notation involving geometrical elements, which had been introduced by Peirce and developed by Spencer-Brown. In this notation, the proof made more human sense than it had done before.

Here, then we have an entirely new pattern of development in mathematics. Automated theorem provers produce a humanly incomprehensible computer proof of a theorem. Human mathematicians then try to make sense of the computer proof by translating it into a humanly more accessible language.

As well as a chapter on automated theorem proving, Corfield includes a further chapter on automated conjecture formation. His investigation of the interactions between computers and mathematics in recent times could obviously be carried further. For example, the introduction and study of fractals would hardly have been possible without computers and computer graphics. Still, it remains true that only a relatively small part of recent mathematics has been involved with computers. The mainstream has been concerned with very abstract mathematics such as Hopf algebras, algebraic topology, and category theory. Corfield's book devotes considerable space to these developments.

Once again Corfield takes Lakatos as his starting point, though this time he focusses on Lakatos' notion of research programme. He is also more critical of Lakatos. In particular, he criticizes Lakatos' rather negative attitude to axiomatization which is expressed in the following passage (Lakatos, c.1959–61, p. 68):

> Axiomatization is a big turning point in the life of a theory, and its importance surpasses its impact on proofs; but its impact on proofs is immense in itself. While in an informal theory there really are unlimited possibilities for introducing more and more terms, more and more hitherto hidden axioms, more and more hitherto hidden rules in the form of new so-called 'obvious' insights, in a formalized theory imagination is tied down to a poor recursive set of axioms and some scanty rules.

Corfield's criticism is as follows (2003, pp. 151–2):

> Lakatos … was led to the mistaken position that the development of modern mathematics lacks much of the freedom and excitement of earlier times. This … prevented him from observing an important part of the dialectical process involved in the development of the mathematics of the twentieth century. I shall suggest instead that the appropriate use of rigorous definition and axiomatization has not acted as a hobble on the creativity of mathematicians, but rather as an invaluable tool in the forging of new mathematical theories and the extension of old ones.

Corfield mentions the Eilenberg–Steenrod axiomatization of homology in 1945 as an example of the fruitful use of axiomatization (2003, p. 186).

Corfield tries to apply Lakatos' notion of research programme to the development of recent mathematics, but he meets with some difficulties. One example he considers was an episode in 1942–5 (2003, p. 184): 'where five mathematicians produced four independent papers continuing an idea of Hopf's in precisely the same direction'. These mathematicians had very different backgrounds, and as Corfield goes on to say (2003, p. 185):

> We have five mathematicians with their different backgrounds and knowledge of a variety of areas of mathematics involved for a brief time in the attempt to achieve similar ends, yet making use of dissimilar means. To count these mathematicians as working within the same research programme would force us to include within the heuristic of that programme a wide variety of favoured means originating from a considerable part of the totality of mathematics. When mathematicians from a variety of backgrounds work temporarily on the same project, the correct attribution of the heuristic becomes a far from simple matter.

This is part of a more general problem which arises from the fact that recent mathematicians (Corfield, 2003, p. 203) 'have increasingly seen fit to relate apparently unconnected theories. Indeed it is the success of the structuralist tradition to have allowed for an accelerated interweaving of research. … Thus, mathematics today progresses in ways in which it could not have progressed a hundred years ago'. As a result of this (Corfield, 2003, p. 200):

Mathematics starts to look like a tangled net and it becomes much less appropriate for a historian to offer a sequential narrative.

Despite these difficulties, Corfield goes on to give an analysis of concepts of recent higher-dimensional algebra, such as categories, categorification, k-tuply monoidal n-categories. On page 257, he makes an observation about 'quasi-triangular Hopf 2-algebras, whose representations form a braided monoidal 2-category'. It has to be said that most other philosophers of mathematics have not followed Corfield into considering these obscure areas, but it does seem to me that Corfield has posed a problem which should be tackled by the philosophy of mathematics. An important change does seem to have occurred which makes recent mathematics significantly different from the mathematics of earlier periods. I would date this change from about the 1940s which saw the introduction of Hopf algebras and of the Eilenberg–Steenrod axiomatic approach to algebraic topology. This raises a number of questions. How can we characterize the difference between mathematics since 1940 and the mathematics of earlier historical periods? What brought about this change in the character of mathematics and what is its significance?

4.4 Kvasz (2008)

The title of Kvasz's book: *Patterns of Change. Linguistic Innovations in the Development of Classical Mathematics* shows two things about its contents. First of all, Kvasz aims to discover patterns of change in the development of mathematics. This is a project typical of the historical approach to philosophy of mathematics. Secondly, however, the title shows that the focus is going to be on language. This marks a departure from the earlier attempts to discover patterns of change in mathematics by considering those already discovered in science and seeing whether, perhaps in a modified form, they held for mathematics as well. This approach from science was of course adopted by Lakatos who modified Popper 'conjectures and refutations' for science to become 'proofs and refutations' for mathematics. It was also adopted in the debate as to whether Kuhnian revolutions in science occurred in mathematics as well. Kvasz, however, introduces a new approach which involves the consideration of language rather than science.

Kvasz's idea is (2008, p. 7) that 'we can *interpret the development of mathematics as a sequence of linguistic innovations*' (Italics in original). As a result of his linguistic approach, Kvasz draws more on ideas from general analytic philosophy than from philosophy of science. More specifically he makes use of the classic works of Frege and the early Wittgenstein. However, Kvasz develops the ideas of Frege and the early Wittgenstein in a number of novel ways. Perhaps most importantly he introduces a historical dimension to the study of language. Both Frege and the early Wittgenstein treat language as timeless, but Kvasz, by contrast, focusses on the historical changes by which an older language can develop into a stronger, richer new language which has greater expressive power. He puts the point as follows (2008, p. 7):

> We will try to understand the language of mathematics historically, i.e., not as some ideal calculus that exists outside of time. Instead we will try to reconstruct the formal languages of a particular period in history and to understand the changes of language which occurred during the transition from one historical period to the next. Thus we will interpret changes in mathematics as *changes* of the language of mathematics.

Kvasz then argues that his linguistic approach has several advantages. As he says (p. 7):

> If we approach a particular change in mathematics from the linguistic point of view, we have at our disposal two languages – the language before and the language after the change. These languages are better accessible to analysis than the psychological act of the discovery or the heuristics that led to it. A language has many objective aspects that can be studied and analysed.

Kvasz makes a successful use of his linguistic approach, and, in his book formulates three novel patterns of change in the development of mathematics. These are (1) *re-codings*, (2) *relativizations*, and (3) *reformulations*. Here, I will discuss in detail only one of the three: relativizations.

Kvasz takes some ideas from Wittgenstein's *Tractatus* as the starting point of his development of the concept of relativization. In the *Tractatus*, Wittgenstein presents the picture theory of language. Regarding pictures, he makes the following important observation (Wittgenstein, 1921, p. 17):

> 2.172 A picture cannot, however, depict its pictorial form: it displays it.

On the picture theory of language, then, language must have a form which is displayed in the language but cannot be expressed in the language. To Kvasz, this suggested a way in which an initial language L_1 say could be transformed historically into a more powerful language L_2. We consider the form of the language of L_1 which cannot be expressed within L_1 itself. However, by adding this form to L_1 we create a new language L_2 which has more expressive power than L_1. The creation of more powerful languages in this manner does indeed, according to Kvasz, occur frequently in the development of mathematics. It is what he calls relativization.

Kvasz first example of a relativization concerns the theory of perspective developed by the Italian painters of the Renaissance. This theory was of course a mathematical theory and its language had what Kvasz calls 'the perspectivist form'. Perspective theory was developed by Desargues into projective geometry. Kvasz comments on the change as follows (2008, pp. 118–19):

> The main achievement of the previous (perspectivist) form was a faithful representation of the theme of the painting from a particular point of view. Nevertheless, the point of view itself was not represented in the painting. It was the point out of which we had to look at the painting, and as such it was unrepresented. The projective form brought a radical change with respect to the point of view: the point of view became explicitly represented in the painting.

In fact, 'the point of view' in the theory of perspective becomes 'the centre of projection' in projective geometry, and so is represented as a point within the geometry rather than being something outside. Projective geometry went beyond earlier forms of geometry in many respects. One of these was the introduction of the 'line at infinity' (corresponding to the horizon in the theory of perspective). This enabled the infinitely distant to be treated in a new fashion. As Kvasz says (2008, p. 122): '*Desargues found a way to give to the term infinity a clear unambiguous and verifiable meaning*' (Italics in original).

Kvasz goes on to analyse, using his concept of relativization, the emergence of ever more complicated forms of synthetic geometry – including most notably non-Euclidean geometry. This approach casts new light on some of the puzzling features of the discovery of non-Euclidean geometry.

Kvasz's treatment of relativization is a good illustration of an impressive feature of his book, namely the wealth of detailed historical case studies from the history of mathematics with which he illustrates his three patterns of change. Of course, not everyone will agree with all the aspects of this treatment, but it leaves little doubt that the three patterns described by Kvasz are indeed patterns which have characteristically recurred in the development of mathematics. The concepts of recoding and relativization are quite novel and original, and so must be considered as constituting a real and substantial contribution to history and philosophy of mathematics. Kvasz has made genuine progress with his linguistic approach. One reason for this situation, in my view, is that mathematics is well suited to the linguistic approach.

Consider a big innovation in mathematics such as the development of calculus. This brought a lot of linguistic changes. These included the introduction of signs for differentials such as dy/dx, and for integrals, such as $\int$. The results of the new calculus were expressed in formulae which were quite different from those of previous mathematics and would have been as incomprehensible to earlier mathematicians as hieroglyphics were to Egyptians living in the eighteenth century. The introduction of the calculus was a re-coding in Kvasz's sense. This example, and others like it, indicates that mathematics is a good place to look for those who want to study how significant changes in language occur. The language of everyday life does indeed change, but at a slow rate which can take centuries. In a few decades mathematicians can develop strikingly new languages. Thus, mathematics is a kind of artificial laboratory in which large linguistic changes can be observed and studied with ease. The importance of mathematics for the study of linguistic change shows that Kvasz's results should be of interest not just to philosophers of mathematics, but also to those with a general interest in the philosophy of language. Kvasz's book was the winner of the first Fernando Gil prize, which was awarded in 2010.

As well as putting forward his own position, Kvasz includes in his book an interesting discussion of how his investigations relate to his predecessors – Lakatos and the Kuhnians. He classifies Lakatos' method of proofs and refutations as a re-formulation, writing (2008, p. 228):

> Re-formulations play an important role in the formation of mathematical concepts and the search for their appropriate definitions. A classic in this area is the book *Proofs and Refutations* by Imre Lakatos.

Kvasz, however, thinks that Lakatos in all his writings only considered reformulations and did not recognise or analyse either re-codings or relativizations (cf. 2008, p. 248, last paragraph).

Up to a point this is quite fair. Lakatos never claimed that *Proofs and Refutations* was the universal pattern of change in mathematics. On the contrary, he stated that it had only been discovered in the 1840s, thereby suggesting that other researchers might be able to discover new patterns of change in mathematics. However, Kvasz discussion of Lakatos might be qualified by two considerations.

First of all, while the method of proofs and refutations may indeed be a reformulation in Kvasz's sense, it is more than that. Lakatos does describe the way that the meaning of 'polyhedron' changed in the course of the historical development, and this is a linguistic change. However, he describes also the formation of a conjecture, the invention of a proof, the discovery of counter-examples, the analysis of the proof into lemmas, the alteration of the conjecture by the incorporation of a proof-generated lemma, and so on. These matters are all important but are not changes in language. Linguistic changes are indeed an important aspect of the development of mathematics, but not the only aspect.

Secondly, in arguing that Lakatos never introduced the concept of relativization, Kvasz says (2008, p. 246):

> *Lakatos' method can only be adopted in cases where the form of language is not changing.* (Italics in original)

If by Lakatos' method is meant the method of proofs and refutations, this is true, but, if we interpret Lakatos' method more broadly as his general approach to philosophy of mathematics, then it would allow, and even encourage, quite drastic linguistic changes. Indeed, Lakatos says (1963–4, p. 324, 1976, p. 93):

> As knowledge grows, languages change. ... The growth of knowledge cannot be modelled in any given language.

Kvasz is not sympathetic to Kuhn, and criticizes him as follows (2008, p. 243):

> The main point that can be raised against Kuhn's theory on the basis of our analyses is that his notion of scientific revolutions includes changes of different kinds. Kuhn includes among revolutions *re-codings*, which he analyses on the example of the Copernican revolution, as well as *relativizations*, an example of which is the oxidation theory of combustion. Therefore his categories, such as paradigm, anomaly, crisis, and revolution that he obtained from the analysis of such heterogeneous material, are rather rough and unspecific.

The Kuhnians, whom we discussed earlier, could reply with a *tu quoque*. As Kvasz records (2008, pp. 240–1), these Kuhnians classified some re-codings (the creation of set theory, the discovery of the calculus, and the discovery of the predicate calculus) as revolutions in mathematics. However, other re-codings, such as the transition from the differential and integral calculus to iterative geometry have never been regarded as revolutions. The situation is just the same with relativizations. Some relativizations such as the change from the theory of perspective to projective geometry have never been regarded as revolutions, while others such as the discovery of non-Euclidean geometry are standardly regarded as revolutions in mathematics. So, one could say that the categories of re-coding and relativization 'obtained from the analysis of such heterogenous material, are rather rough and unspecific'.

This is not, however, how I would like to analyse the situation. It seems to me that Kuhn's categories and those of Kvasz are both valuable but are concerned with different aspects of the problem of scientific and mathematical change. The problem therefore is to integrate these categories. Perhaps a way of doing so could be found in Wittgenstein's later theory of meaning. The meaning of a word in a mathematical language could be regarded as its use by the mathematical community (a Kuhnian category). In the spirit of the later Wittgenstein, I would take the mathematical community to include not just research mathematicians, but those who use mathematics in a variety of applications. These applications are surely language games in the sense of the later Wittgenstein.

4.5 Grosholz (2016)

Emily Grosholz's 2016 book: *Starry Reckoning: Reference and Analysis in Mathematics and Cosmology* was the winner of the 2017 Fernando Gil prize. As the title indicates, it deals with issues in both philosophy of mathematics and philosophy of science, but here I will discuss only those parts concerned with philosophy of mathematics. Grosholz states clearly at the beginning that she is adopting an historical approach and traces this back to Imre Lakatos (2016, p. vi):

> I am happy to note that philosophy of mathematics in the early 21st century is undergoing a long-awaited and long-overdue sea change. Philosophers of mathematics are turning to a serious study of the *history* of mathematics, logic, and philosophy. . . . They are trying to give an account of how mathematical knowledge grows and how problems are solved, with the dawning insight that serious problem-solving is by nature ampliative. . . . In fact, this change had been in preparation for decades. Imre Lakatos' celebrated book *Proofs and Refutations; The Logic of Mathematical Discovery* was published in 1976, and inspired many young philosophers to think about mathematics in a new way.

In her 2016 book, Grosholz takes as central the concepts of *reference* and *analysis*. As she says (2016, p. xiv):

> In this book, my focus is on the following patterns of reasoning. I argue that productive mathematical . . . research takes place when discourse whose main intent is to establish and clarify reference is yoked with discourse whose main intent is analysis.

Now here a clarification is needed because the term 'analysis' might be understood as meaning 'logical analysis' as it is in analytic philosophy. However, Grosholz uses it in another sense, as she goes on to explain (2016, p. xiv):

> I borrow the term analysis here from Leibniz: it means both the search for conditions of solvability of problems, and the search for conditions of intelligibility of things.

This Leibnizian notion of analysis has close connections with an historical approach, because we can only render some concepts intelligible by considering how they developed historically. Grosholz gives a very nice illustration of this with the example of the relation between the sine and cosine functions and the circle (See Figure 4,[11] reproduced from Grosholz, 2016, p. 13).

Grosholz comments on this as follows (2016 p. 12):

---

[11] This is Figure 1 in the original text.

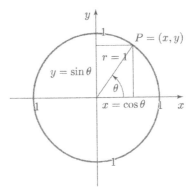

Figure 4 Sine and cosine functions in the *circle*

We cannot deduce from Euclid's definition the fact that the sine and cosine functions can be generated in terms of the circle. They are 'contained in' the concept of circle for Euler, but not for Euclid. How did they show up there, right in the middle, plain as day? The answer is given by Leibniz's notion of analysis, for Descartes and Leibniz himself discovered new conditions of intelligibility for the circle by bringing it into novel relation with algebra and arithmetic. We need the Pythagorean theorem (which is about triangles). Cartesian geometry (which is not only about geometry but also about the algebra of arithmetic), the infinitesimal calculus, the completion of the rationals that we call the reals and the notion of a transcendental function. Once we have embedded the circle in Cartesian geometry, algebra, arithmetic and infinitesimal analysis, and Euler has developed the modern notion of a function for us, we can see that the functions sine and cosine can be 'read off' the circle.

This passage contains a number of Grosholz's favourite themes. The historical development which forms part of her Leibnizian notion of analysis, takes place through the bringing together of different domains. Cartesian geometry, for example, was created by bringing together geometry and algebra. Although Grosholz regards it as fruitful to bring different domains together, she also regards it as a mistake to try to reduce one domain to the other. The domains should retain some autonomy while interacting, and this in turn means that a concept in the intersection of two domains has some degree of ambiguity, taking its meaning partly from its use in one domain and partly from its use in the other. This ambiguity is not, however, a defect, but rather a productive ambiguity because it stimulates further development.

An example of these themes of domain interaction and the resulting productive ambiguity is provided by (2016, p. xvi) 'the disparity between the referential discourse of arithmetic and number theory and the analytic discourse of logic'.

Grosholz' discussion of this example leads to a general account of the relation between logic and mathematics. She regards the logicists as wrong in their claim that mathematics can be reduced to logic. In fact, she thinks that (2016, p. vi)

> Mathematical logic ... is not an over-discourse that should supplant the others, but one of many, which can be integrated with other mathematical discourses in a variety of fruitful ways.

This point of view is developed in detail through a number of examples.

The first concerns Enderton's account of logic in his 1972 book: *A Mathematical Introduction to Logic*. Grosholz comments (2016, p. 86):

> For Enderton, the 'intended structure' U = (N, 0, S, <, +, . , E) has already been assimilated to a logician's discourse in which the numbers are represented as the initial element 0 and successive iterations of the function successor, S. This mode of representation was developed by logicians to subject arithmetic to logical analysis, but it is never employed in the classroom to teach arithmetic, or in articles in journals devoted to number theory. When logicians need to use the natural numbers as indices, or to compute number theoretic facts like the prime decomposition of a large number, they make use of Indo-Arabic/Cartesian notation without really mentioning their own departure from the formalism they are supposed to be using.

This is the difference between the analytic discourse of logic in which numbers appear as 0, S0, SS0, ..., and the referential discourse of arithmetic and number theory in which numbers assume their more familiar decimal form such as 12,964. There is a disparity between these two discourses, so that logicism fails in its attempt to reduce arithmetic to logic. Yet the two discourses can be brought fruitfully together, as Grosholz illustrates by a very interesting and insightful discussion of Gödel's proof of his incompleteness theorems.

Gödel's original proof concerned Russell and Whitehead's logicist system *Principia Mathematica*. Now within this system the natural numbers are defined in purely logical terms and appear as 0, f0, ff0, fff0, ... These belong to the analytic discourse of logic. However, Gödel's first step in his proof is to assign a numbering to the terms and formulas of *Principia Mathematica*. In fact (Gödel, 1931, p. 601) he assigns '0' the number 1, and 'f' the number 3. These 'Gödel numbers' belong to the referential discourse of arithmetic and number theory. The proof brings these two discourses together and moves from one sense of number to the other. As Grosholz puts it (2016, pp. 88–9):

> I am trying to show that to carry out his proof, he must use modes of representation that lend themselves to logical analysis (Russell's notation) but not to computating or referring, and other modes of representation that

lend themselves to successful reference (Indo-Arabic/Cartesian notation). He must use disparate registers of the formal languages available to him, combine them, and exploit their ambiguity.

It will be remembered that David Corfield (see Section 4.3) challenged philosophers of mathematics to consider examples drawn from contemporary mathematics. Emily Grosholz responds to this challenged by giving an analysis of one of the most striking results of recent mathematics, namely Wiles' proof in 1995 of Fermat's Last Theorem. Wiles' proof is in fact very suited to Grosholz' approach since it puts together different areas of mathematics. As Grosholz says (2016, p. 96):

> In the case of number theory, the referents are integers and rational numbers in one sense and additionally, in a broader sense given the problem reduction at the heart of Wiles' proof, modular forms and elliptic curves.

Wiles' proof affords another good example of Grosholz' theory of the relation between logic and the rest of mathematics. Wiles' proof makes use of cohomology theory and Grothendieck's notion of successive universes. Now it can be proved within Grothendieck's universes that ZFC is consistent. This shows that they go beyond ZFC. Yet a mathematical theorem is normally accepted as valid if it can be proved within ZFC. If a theorem is only provable within a system stronger than ZFC, can it still be accepted as valid? This question has raised the problem of whether Wiles' proof can be carried out within some weaker system. Grosholz mentions the work of Angus Mcintyre who has asked what the proof might look like if confined within first-order Peano Arithmetic. His investigations have led to results of a kind different from those of Wiles. As Grosholz says (2016, p. 100):

> Mathematicians like Wiles, Ribet and Mazur posit the big structures to set their problem in the best conceptual framework possible, so they can see how to solve the problem and then how to generalize the result; model theorists like Mcintyre break the big structures into smaller approximations in order to solve different kinds of problems.

4.6 Cellucci 2022

Carlo Cellucci's 2022 book has the title: *The Making of Mathematics. Heuristic Philosophy of Mathematics*. Its aim is to present a heuristic philosophy of mathematics in which the main focus is on how mathematics is created, that is to say on questions of discovery in mathematics and the development of mathematics. As to the origin of this type of philosophy of mathematics, Cellucci is quite clear, for he writes (2022, p. 60):

> The original formulation of heuristic philosophy of mathematics can be credited to Lakatos's Ph.D. dissertation.

This is the dissertation completed in 1961 with the title: *Essays in the Logic of Mathematical Discovery.*

At the core of Cellucci's heuristic philosophy of mathematics is what he calls 'the analytic method' and whose introduction he attributes to Plato. Here is a summary which Cellucci gives of this method (2022, p. 60):

> The method of mathematics is the analytic method. The latter is the method according to which to solve a problem, one looks for some hypothesis that is a sufficient condition for solving the problem, namely such that a solution to the problem can be deduced from the hypothesis. The hypothesis is obtained from the problem and possibly other data, by some non-deductive rule, and must be plausible, namely such that the arguments for the hypothesis are stronger than the arguments against it, on the basis of experience. But the hypothesis is in turn a problem that must be solved, and is solved in the same way.

Note that in this account 'the hypothesis is obtained . . . by some non-deductive rule'. Now Lakatos did not use non-deductive rules and he is criticized by Cellucci for this omission. In this respect, Cellucci is quite typical of the followers of Lakatos who generally develop some of Lakatos' ideas while criticizing others. Cellucci writes (2022, p. 64):

> Lakatos has no methodology qua logic of discovery. This does not invalidate the claim that the original formulation of heuristic philosophy of mathematics can be credited to Lakatos. But it means that, with respect to heuristic philosophy of mathematics, Lakatos is a sort of 'non-playing captain', namely a captain who is not in the field when the game takes place.

Of course, Lakatos might well have objected to this criticism. He writes (1976, p. 127):

> The method of proofs and refutations is a very general heuristic pattern of mathematical discovery.

So, he could have claimed that 'proofs and refutations' is a 'methodology, qua logic of discovery' for mathematics. However, Cellucci points out that the method of proofs and refutations begins with the formulation of a naïve conjecture, but as he goes on to say (2022, p. 62):

> Lakatos's method . . . does not account for how to arrive at the naïve conjecture.

In fact, Lakatos appeals to 'trial and error' which Cellucci does not regard as acceptable as part of a logic of discovery. Similar points apply to the formulation of the proof and the discovery of counterexamples. Lakatos did indeed formulate a heuristic pattern, but this heuristic pattern is not precise enough, according to Cellucci, to count as a logic of discovery. It needs to be supplemented by the rules of a non-deductive logic. As Cellucci says (2022, p. 80):

> To give a rational explanation of mathematical creativity, non-deductive logic is needed. Heuristic philosophy of mathematics starts from here ... in the analytic method on which heuristic philosophy of mathematics is based, non-deductive arguments generate the hypotheses for solving problems. While deductive arguments do not generate new ideas, non-deductive arguments generate new ideas by providing hypotheses for solving problems, whose plausibility is then investigated. Mathematical creativity consists in this. Thus, mathematical creativity is not based on illumination, of which no rational account can be given, but on non-deductive logic, which is genuinely logic.

In Chapter 7 of his book, Cellucci formulates some rules of this non-deductive logic. These involve induction, analogy, metaphor, metonymy, generalization, and specialization. I will illustrate these rules by considering one example, induction from multiple cases or IMC (Section 7.4). Cellucci gives an example from the work of the mathematician Bachet, which was published in 1621, and writes as follows (2022, p. 198):

> Let us consider the problem: Is every natural number either a square or the sum of two, three, or four squares?
>   To solve the problem, Bachet observes that $1 = 1$, $2 = 1 + 1$, $3 = 1 + 1 + 1$, $4 = 4$, $5 = 1 + 4$, $6 = 1 + 4 + 1$, $7 = 1 + 1 + 1 + 4$, $8 = 4 + 4, 9 = 9, 10 = 1 + 9, \ldots, 325 = 1 + 324$, so every natural number up to 325 is either a square, or the sum of two, three, or four squares. From this, by (IMC), Bachet arrives at the hypothesis: 'Every number is either a square, or the sum of two, three, or four squares'.

Cellucci is undoubtedly correct that this type of induction has been widely used in mathematics – particularly in number theory. As he points out, it was employed by Euler and Fermat. Moreover, using computers, the number of instances on which the induction is based can be vastly increased. The Riemann hypothesis is the conjecture that the Riemann zeta function has its zeros only at negative even integers and complex numbers with real part ½. In 2020, it was verified by computer for the first 12,363,153,473,138 zeros. This is a much larger base than 325 from which to make inductions.

In fact, Euler did discover his conjecture that $V - E + F = 2$ by an induction of this kind from the then standard examples of polyhedra. So Cellucci is also right that a non-deductive rule of discovery can supplement Lakatos' method of proofs and refutations. Lakatos would have been reluctant to admit the existence of such a rule in 1963–4, because he was, at that time, an enthusiastic follower of Popper and would have accepted Popper's claim that induction does not exist.

Let us now turn to another feature of Cellucci's philosophy of mathematics, namely his treatment of the axiomatic method. As we saw in Section 4.3, Corfield disagreed with a criticism of the axiomatic method by Lakatos that the formulation of axioms might slow down mathematical progress, and argued that in the post-1940 period, the formulation of axioms, such as the Eilenberg–Steenrod axioms for homology, has stimulated the advance of mathematics. In this debate, Cellucci tends to side with Lakatos rather than Corfield. Cellucci discusses the formal axiomatic method, saying (2022, p. 182):

> The formal axiomatic method arose gradually from the second half of the nineteenth century to the early decades of the twentieth century, but is best known from Hilbert's formulation.

He characterizes it as follows (2022, p. 181):

> The choice of the axioms is arbitrary, subject only to the conditions that the axioms must be consistent. The primitive terms occurring in the axioms can be thought of in any way one likes.

Cellucci argues that axiomatic demonstration does not provide an account of how mathematicians discover results (2022, p. 264):

> For, according to it, results are discovered starting from the principles and deducing consequences from them. But, as a matter of fact, most mathematical results are not discovered in that way.

So, at best, (2022, p. 181):

> The purpose of the formal axiomatic method ... is not to obtain new knowledge, but only to present, justify, and teach already acquired knowledge.

However, Cellucci develops quite a number of further criticisms of the axiomatic method. First of all, he stresses (2022, p. 217) that the axiomatic view of theories is shown to be inadequate by Gödel's incompleteness theorems. Even a strong defender of the axiomatic method can hardly deny the justice of this, but such a defender might say that, although Gödel's incompleteness theorems do indeed show that there are limitations to the axiomatic method, it is still useful in some areas of mathematics. For example, it could be claimed that Kolmogorov's axiomatization of probability has proved very useful for the development of the mathematical theory of probability.

Cellucci, however, sees the formal axiomatic method as leading to 'the sclerosis of research'. He explains this in a passage which gives an interesting, but not widely known, quotation from Frege (Cellucci, 2022, p. 191):

Frege says that, 'where a tree lives and grows, it must be soft and succulent', and, 'when all that was green has turned into wood, the tree ceases to grow' (Frege, 1980, p. 33). Similarly, we can say that, where mathematics lives and grows, it must be flexible and fecund, and, when mathematics is organized and presented by the formal axiomatic method, it ceases to grow. To the lignification of the tree, which is the end of its growth, there corresponds the sclerosis of research produced by the formal axiomatic method, which is the end of its flexibility and fecundity.

Cellucci argues that his analytic method gives a better account of mathematics than the axiomatic method. Indeed, following the analytic method a mathematician dealing with some problem considers a hypothesis which might solve it. However, he or she is not limited to considering only hypotheses which are consistent with some set of axioms. So, the analytic method allows more freedom to the researcher and hence might lead to superior growth in mathematical knowledge.

On the other hand, the analytic method does not allow complete freedom to the researcher, for, as Cellucci says (2022, p. 60):

The hypothesis is obtained from the problem and possibly other data, by some non-deductive rule, and must be plausible, namely such that the arguments for the hypothesis are stronger than the arguments against it, on the basis of experience.

These desirable constraints are removed by the formal axiomatic method, and this has a number of negative consequences, including the separation of mathematics from physics, which impacts badly on the quality of mathematical research. As Cellucci says (2022, pp. 190–1):

The separation leads to a fragmentation of mathematics. Indeed by permitting to consider arbitrary axioms with the only condition of consistency, it produces an uncontrolled proliferation of theories. ... The separation leads mathematicians to create artificial systems, entirely without applications.

Cellucci also quotes (2022, p. 190) the following from Bourbaki (1950):

Many mathematicians take up quarters in a corner of the domain of mathematics, which they do not intend to leave. Not only do they ignore almost completely what does not concern their special field, but they are unable to understand the language and the terminology used by colleagues who are working in a corner remote from their own.

These passages suggest that the disagreement between Corfield and Cellucci about the axiomatic method may have its origin in a difference in judgement concerning the value of contemporary abstract mathematics. Corfield is an admirer of the new mathematics, whereas Cellucci seems to see most of it as a fragmented set of artificial systems, often entirely without applications.

I have tried here to give a brief account of Cellucci's own philosophy of mathematics, but his book contains, in addition, detailed criticisms of alternative positions in the field. He also applies his approach to a range of problems in the philosophy of mathematics which concern: objects, demonstrations, definitions, diagrams, notations, explanations, beauty in mathematics, and the applicability of mathematics. Cellucci (2022) is thus a substantial work and, as such, is a fitting one with which to end this survey of the legacy of Lakatos.

5 Concluding Remarks

Lakatos was born on 9 November 1922, and so at the time when I am writing this passage (9 February 2023) he would, if still alive, be 100 years old. That is an advanced age, but not an impossible one. In fact, my brother's mother-in-law, Muriel Jones, was born on 11 April 1922, more than six months before Lakatos, and she is still alive today.

So, what if Lakatos had lived till the present, how would this have affected the developments described in Sections 3 and 4? This is an interesting question, but one which I don't think can be answered in a satisfactory fashion. If Lakatos had lived longer, he would undoubtedly have returned to the philosophy of mathematics and made more contributions to this subject; but we have no means of knowing what these contributions would have been. He would also have criticized the contributions of others and we have again no idea about what changes this would have introduced. For these reasons I prefer to consider another imaginary scenario.

Suppose that by the incantation of a Latin spell, we were able to raise the ghost of Lakatos from the dead, and to allow this ghost to study the developments described in Sections 3 and 4, what judgement would the ghost give on them? I have already observed that Lakatos' legacy contains a great number of criticisms of Lakatos' ideas, for, typically, someone who admired some idea of Lakatos and tried to develop it, would also be critical of other Lakatosian ideas. Lakatos, however, had a great love of controversy and was always ready to reply to any criticisms of his work. I don't therefore think that his ghost would be disturbed by these criticisms but would rather be prepared to answer them in no uncertain terms. As well as controversy, Lakatos always loved new ideas and novel developments. So, his ghost would find much to please him in the variety of different approaches to be found in Lakatos' legacy. The discussion of whether there are revolutions in mathematics was something quite new and led to the formulation of a number of novel positions. Another novel project was that of trying to reconstruct philosophical ideas about mathematics in the seventeenth century and to examine the influence of these ideas on the mathematical practice of the time. This investigation gave some surprising results. David Corfield was the first person to stress that philosophy of mathematics should take account of contemporary non-foundational mathematics. He made a striking contribution to this novel programme, and Emily Grosholz contributed to it as well with her analysis of Wiles' 1995 proof of Fermat's last theorem. Ladislav Kvasz had the novel idea of studying the development of mathematics as a sequence of linguistic innovations, and this approach led him to formulate a number of new patterns of change in mathematics, such as re-codings and relativizations. Carlo Cellucci elaborated the 'analytic method' introduced by Plato to give a full account of mathematics. There is little doubt in my mind that Lakatos' ghost would find all this very interesting and pleasing and would feel considerable satisfaction that such a rich crop of new ideas and novel developments had grown out of his own innovations in the philosophy of mathematics.

We can consider both Lakatos' own contribution and his legacy as part of a research programme whose aim is to develop a historical approach in the philosophy of mathematics. This could be called *the historical programme in the philosophy of mathematics*. Lakatos himself first used the term 'research programme' in his 1968 paper 'Changes in the problem of inductive logic'.[12] The paper begins (1968, p. 128):

> A successful research programme bustles with activity. There are always dozens of puzzles to be solved and technical questions to be answered; even if *some* of these – inevitably – are the programme's own creation.

The historical programme in the philosophy of mathematics has certainly bustled with activity since its initiation by Lakatos in 1963–4. Lakatos is right that a research programme, as it progresses, creates some of its own problems, and it is interesting therefore to consider what problems have emerged from the historical programme in the philosophy of mathematics. I will here mention three.

(1) There is the problem of whether a historical philosophy of mathematics should include Kuhn's social concept of the mathematical community and a social analysis of this community, or whether, as Lakatos himself insisted, it should concentrate on the objective ideas of mathematics and their development. I favour the integration of Kuhn's concepts with Lakatos' concept of research programme, since I see these concepts as different and compatible rather than contradictory. However, Kvasz in his 2008 has argued for a more Lakatosian and anti-Kuhnian position.

---

[12] This was the paper on which Lakatos was working during my first year (1966–7) as his PhD student.

(2) Modern logic was originally developed as part of the foundational programmes of logicism and formalism. These programmes received a severe blow from Gödel's incompleteness theorems, and this raises the problem of what account we should now give of logic. Emily Grosholz has provided such an account in her 2016. She thinks that contemporary mathematical logic should be regarded as a mathematical discourse like any other, which can be integrated with other mathematical discourses in a variety of fruitful ways. But is that all there is to modern logic? Does modern logic not retain some foundational significance, albeit of a weaker form than was once hoped? Is mathematical logic not closer to philosophy than some other branches of mathematics? Other questions about logic are raised by Carlo Cellucci's attempt in his 2022 to develop a non-deductive logic. Some of the rules he proposes for such a logic, such as IMC (induction from multiple cases) do seem both plausible and applicable to mathematics. But how far can one go in developing a logic of mathematical discovery, and how will such a logic relate to deductive logic?

(3) There is the problem of the significance of the axiomatic method in mathematics. Is the axiomatic method helpful in stimulating the development of new mathematics, as David Corfield claims? Or is the axiomatic method a baleful influence on mathematics which would be better replaced by the analytic method, as Carlo Cellucci claims? It is interesting to note that there is some ambiguity on this in Lakatos' own writings. The passage criticized by Corfield in which Lakatos laments the restrictions which axiomatization places on the development of mathematics was actually written in the period 1959–61. In Lakatos' last paper on philosophy of mathematics written in 1973, the year before he died, he suggests that Euclid's axioms might be considered as the hard core of a research programme. Since research programmes can, after all, be progressive, this suggests that axiomatization might lead to further mathematical development, though perhaps of a different kind from the development in the informal pre-axiomatic period. If Kuhnian terms are allowed, we might see a successful axiomatization as constituting a paradigm and so initiating a period of 'normal' mathematics in which progress is made by a mathematical community who all accept the common framework provided by the axioms. An example of this might be Kolmogorov's successful axiomatization of probability theory in 1933, which, despite some challenges, has provided the framework within which the mathematical theory of probability has been developed successfully up to the present.

That concludes my account of some of the problems which have emerged from the historical programme in the philosophy of mathematics. Now I want to raise a more general question about this programme.

Lakatos always asked whether a research programme should be considered as progressing or degenerating, and so how should we regard the historical programme in the philosophy of mathematics in this respect? It seems to me clear that the programme has been progressive from its initiation to the present. In the light of all the interesting new ideas which have emerged from the programme, it would be difficult to make any other judgement. We could, however, ask whether this is likely to continue, or whether the programme has reached its limits and might even be starting to degenerate. Of course, it is always hard to say what the future holds, but it seems to me that the programme shows every sign of continuing to be progressive. Almost all the approaches which have been introduced could be developed further and the programme has thrown up a number of problems which are far from being resolved. I have just described three such problems. Moreover, there are some problems in the philosophy of mathematics which have not been widely considered by those working on the historical programme, but which could be included within it with potentially fruitful consequences. I will now describe these problems.

From the time of the ancient Greeks, philosophy of mathematics has been concerned with some general philosophical issues of epistemology and ontology. The question 'what is the nature of mathematical knowledge?' has always been a tricky one for theories of epistemology. The questions 'do mathematical entities such as numbers exist? And, if so, what kind of existence do they have?' are difficult ones for theories of ontology. In my account of Lakatos' legacy, however, these epistemological and ontological questions have not appeared. Admittedly, this is partly the result of selection bias. There have been some followers of Lakatos who have written extensively on some of these general philosophical issues. A good example is Gianluigi Oliveri. As the title of his 2007 book: *A Realist Philosophy of Mathematics* shows, it is largely concerned with mathematical ontology. Then again, there are some discussions of these general philosophical issues in the texts I consider, for example in Cellucci (2022), but I have not included these discussions in my brief accounts of the texts.

Yet, even allowing for these selection biases, I think it is still true that these general philosophical questions of epistemology and ontology have not been widely considered by those working on the historical programme in the philosophy of mathematics. This is not surprising. The historical approach introduced some new questions which had not been previously considered by philosophers of mathematics. These were questions of how mathematics develops, what are the patterns of change in mathematics, what heuristics can be used for mathematical discovery, and so on. It was natural for those who first worked on the programme to focus on the new questions which it introduced. Now, however, there seems to be a case of trying to extend the programme to include some of the more traditional philosophical questions concerned with mathematics. The historical programme could introduce a new and fruitful approach to these questions, and so generate progress in the field.

Suppose, for example, someone is trying to develop a theory of the nature of mathematical knowledge, and a theory of the existence of mathematical objects. It could be very helpful to try to test out these theories with a historical case study. Consider the situation in 1706. Should the differential and integral calculus be considered part of mathematical knowledge at that time? And what about the existence of infinitesimals?[13] Would it have been reasonable to regard them as existing in 1706? Leibniz at that time considered them to be *fictions bien fondées* (well-founded fictions), a phrase which has some resonances with contemporary theories of mathematical objects. The example of infinitesimals shows that the view, sometimes expressed, that questions about the existence of mathematical entities are not relevant to mathematical practice, is false. At least it was false in 1706.

All these considerations suggest that the historical programme in the philosophy of mathematics is still progressive and could be further developed fruitfully.

---

[13] Oliveri includes an historical case-study of infinitesimals in his book. See Oliveri, 2007, pp. 11–17.

References

Bourbaki, Nicholas (1950), The Architecture of Mathematics, *The American Mathematical Monthly*, 57, pp. 221–32.

Cellucci, Carlo (2022), *The Making of Mathematics: Heuristic Philosophy of Mathematics*, Springer.

Corfield, David (2003), *Towards a Philosophy of Real Mathematics*, Cambridge University Press.

Crowe, Michael (1975), Ten 'Laws' Concerning Patterns of Change in the History of Mathematics, *Historia Mathematica*, 2, pp. 161–6. Citations taken from the reprint in Donald Gillies (Ed.), 1992a, pp. 15–20.

Dauben, Joseph (1979), *Georg Cantor: His Mathematics and Philosophy of the Infinite*, Harvard University Press.

Dauben, Joseph (1984), Conceptual Revolutions and the History of Mathematics: Two Studies in the Growth of Knowledge. In Everett Mendelsohn (Ed.), *Transformation and Tradition in the Sciences: Essays in Honor of I. Bernard Cohen*, Cambridge University Press, pp. 81–103. Citations taken from the reprint in Donald Gillies (Ed.), 1992a, pp. 49–71.

Duhem, Pierre (1904–5), *The Aim and Structure of Physical Theory*. English translation by Philip P. Wiener of the second French edition of 1914, Atheneum, 1962.

Dunmore, Caroline (1992), Meta-Level Revolutions in Mathematics. In Donald Gillies (Ed.), *Revolutions in Mathematics*, Oxford University Press, 1992a, pp. 208–25.

Dutilh Novaes, Catarina (2021), *The Dialogical Roots of Deduction: Historical, Cognitive, and Philosophical Perspectives on Reasoning*, Cambridge University Press.

Enderton, Herbert (1972), *A Mathematical Introduction to Logic*, Academic Press.

Euclid (c. 300 BC), *The Elements*. English translation by Sir Thomas L. Heath in Robert Maynard Hutchins (Ed.), *Great Books of the Western World*, Encyclopaedia Britannica, 1952, pp. 1–396.

Frege, Gottlob (1918–19), Logical Investigations. Part I Thoughts. In Brian McGuinness (Ed.), *Gottlob Frege, Collected Papers on Mathematics, Logic, and Philosophy*, Blackwell, 1984, pp. 351–72.

Frege, Gottlob (1980), *Philosophical and Mathematical Correspondence*, Blackwell.

Gillies, Donald (Ed.) (1992a), *Revolutions in Mathematics*, Oxford University Press.

Gillies, Donald (1992b), The Fregean Revolution in Logic. In Donald Gillies (Ed.), *Revolutions in Mathematics*, Oxford University Press, 1992a, pp. 265–305.

Gillies, Donald (2014), Should Philosophers of Mathematics Make Use of Sociology? *Philosophia Mathematica*, 22 (1), pp. 12–34.

Giorello, Giulio (1992), The 'Fine Structure' of Mathematical Revolutions: Metaphysics, Legitimacy, and Rigour. The Case of the Calculus from Newton to Berkeley and Maclaurin. In Donald Gillies (Ed.), *Revolutions in Mathematics*, Oxford University Press, 1992a, pp. 134–68.

Gödel, Kurt (1931), On Formally Undecidable Propositions of Principia Mathematica and Related Systems I. English translation in Jean van Heijenoort (Ed.), *From Frege to Gödel: A Source Book in Mathematical Logic, 1879–1931*, Harvard University Press, pp. 596–616.

Grosholz, Emily (2016), *Starry Reckoning: Reference and Analysis in Mathematics and Cosmology*, Springer.

Guicciardini, Niccolò (2009), *Isaac Newton on Mathematical Certainty and Method*, The MIT Press.

Hallett, Michael (1979), Towards a Theory of Mathematical Research Programmes, *The British Journal for the Philosophy of Science*, 30, I pp. 1–25, and II pp. 135–59.

Koetsier, Teun (1991), *The Philosophy of Imre Lakatos: A Historical Approach*, North-Holland.

Kvasz, Ladislav (2008), *Patterns of Change: Linguistic Innovations in the Development of Classical Mathematics*, Birkhäuser.

Lakatos, Imre (c. 1959–61), What Does a Mathematical Proof Prove? First Published in Lakatos, 1978b, John Worrall and Gregory Currie (Eds.), *Mathematics, Science and Epistemology: Philosophical Papers Volume 2*, Cambridge University Press, pp. 61–9.

Lakatos, Imre (1963–4), Proofs and Refutations, *The British Journal for the Philosophy of Science*, XIV I (53 May 1963), pp. 1–25, II (54 August 1963), pp. 120–39, III (55 November 1963), pp. 221–45, IV (56 February 1964), pp. 296–342.

Lakatos, Imre (1968), Changes in the Problem of Inductive Logic. Reprinted in Lakatos, 1978b, John Worrall and Gregory Currie (Eds.), *Mathematics, Science and Epistemology: Philosophical Papers Volume 2*, Cambridge University Press, pp. 128–200.

Lakatos, Imre (1970), Falsification and the Methodology of Scientific Research Programmes. Reprinted in Lakatos, 1978a, *The Methodology of Scientific Research Programmes: Philosophical Papers Volume 1*, Cambridge University Press, pp. 8–101.

Lakatos, Imre (c. 1973), The Method of Analysis-Synthesis. Part 2 Analysis-Synthesis: How Failed Attempts at Refutation May Be Heuristic Starting Points of Research Programmes. First Published in Lakatos, 1978b, *Mathematics, Science and Epistemology: Philosophical Papers Volume 2*, Cambridge University Press, pp. 93–103.

Lakatos, Imre (1976), *Proofs and Refutations: The Logic of Mathematical Discovery*. John Worrall and Elie Zahar (Eds.), Cambridge University Press.

Lakatos, Imre (1978a), *The Methodology of Scientific Research Programmes: Philosophical Papers Volume 1*, John Worrall and Gregory Currie (Eds.), Cambridge University Press.

Lakatos, Imre (1978b), *Mathematics, Science and Epistemology: Philosophical Papers Volume 2*, John Worrall and Gregory Currie (Eds.), Cambridge University Press.

Larvor, Brendan (1998), *Lakatos: An Introduction*, Routledge.

Mach, Ernst (1883), *The Science of Mechanics: A Critical and Historical Account of Its Development*, 6th American ed., Open Court, 1960.

Mancosu, Paolo (1996), *Philosophy of Mathematics and Mathematical Practice in the Seventeenth Century*, Oxford University Press.

Oliveri, Gianluigi (2007), *A Realist Philosophy of Mathematics*, College.

Popper, Karl (1963), *Conjectures and Refutations: The Growth of Scientific Knowledge*, Routledge & Kegan Paul.

Popper, Karl (1972), *Objective Knowledge: An Evolutionary Approach*, Oxford University Press.

Whewell, William (1837), *History of the Inductive Sciences from the Earliest to the Present Times*, 3 volumes, John Parker.

Whewell, William (1840), *The Philosophy of the Inductive Sciences Founded upon Their History*, 2 volumes, John Parker.

Wittgenstein, Ludwig (1921), *Tractatus Logico-Philosophicus*. English translation by David Francis Pears and Brian Francis McGuinness, Routledge & Kegan Paul, 1963.

Cambridge Elements ≡

The Philosophy of Mathematics

Penelope Rush
University of Tasmania

From the time Penny Rush completed her thesis in the philosophy of mathematics (2005), she has worked continuously on themes around the realism/anti-realism divide and the nature of mathematics. Her edited collection *The Metaphysics of Logic* (Cambridge University Press, 2014), and forthcoming essay 'Metaphysical Optimism' (*Philosophy Supplement*), highlight a particular interest in the idea of reality itself and curiosity and respect as important philosophical methodologies.

Stewart Shapiro
The Ohio State University

Stewart Shapiro is the O'Donnell Professor of Philosophy at The Ohio State University, a Distinguished Visiting Professor at the University of Connecticut, and a Professorial Fellow at the University of Oslo. His major works include *Foundations without Foundationalism* (1991), *Philosophy of Mathematics: Structure and Ontology* (1997), *Vagueness in Context* (2006), and *Varieties of Logic* (2014). He has taught courses in logic, philosophy of mathematics, metaphysics, epistemology, philosophy of religion, Jewish philosophy, social and political philosophy, and medical ethics.

About the Series

This Cambridge Elements series provides an extensive overview of the philosophy of mathematics in its many and varied forms. Distinguished authors will provide an up-to-date summary of the results of current research in their fields and give their own take on what they believe are the most significant debates influencing research, drawing original conclusions.

Cambridge Elements ☰

The Philosophy of Mathematics

Printed in the United States
by Baker & Taylor Publisher Services